The Disruptive Impact of Lethal Autonomous Weapons Systems Diffusion

Challenging the focus on great powers in the international debate, this book explores how rising middle power states are engaging with emerging major military innovations and analyses how this will affect the stability and security of the Indo Pacific.

Presenting a data-based analysis of how middle power actors in the Indo-Pacific are responding to the emergence of military Artificial Intelligence and Killer Robots, the book asserts that continuing to exclude non-great power actors from our thinking in this field enables the dangerous diffusion of Lethal Autonomous Weapon Systems (LAWS) to smaller states and terrorist groups, and demonstrates the disruptive effects of these military innovations on the balance of power in the Indo-Pacific. Offering a detailed analysis of the resource capacities of China, the United States, Singapore and Indonesia, it shows how major military innovation acts as a circuit breaker between competitor states, disrupting the conventional superiority of the dominant hegemonic state and giving a successful adopter a distinct advantage over its opponent.

This book will appeal to researchers, end-users in the military and law enforcement communities, and policymakers. It will also be a valuable resource for researchers interested in strategic stability for the broader Asia-Pacific and the role of middle power states in hegemonic power transition and conflict.

Austin Wyatt is a Research Associate at UNSW, Canberra. His research focuses on autonomous weapons, with a particular emphasis on their disruptive effects in Asia. He was awarded his PhD summa cum laude in 2020 from the Australian Catholic University. Dr Wyatt was a New Colombo Plan Scholar and completed a research internship at KAIST. Dr Wyatt's latest published research includes "Charting Great Power Progress toward a Lethal Autonomous Weapon System Demonstration Point", *Defence Studies*, 20(1), 2020 and "The Revolution of Autonomous Systems and Its Implications for the Arms Trade". In *Research Handbook on the Arms Trade* (Edward Elgar Publishing).

Emerging Technologies, Ethics and International Affairs
Series Editors: Steven Barela, Jai C. Galliott,
Avery Plaw, Katina Michael

This series examines the crucial ethical, legal and public policy questions arising from or exacerbated by the design, development and eventual adoption of new technologies across all related fields, from education and engineering to medicine and military affairs. The books revolve around two key themes:

- Moral issues in research, engineering and design
- Ethical, legal and political/policy issues in the use and regulation of Technology

This series encourages submission of cutting-edge research monographs and edited collections with a particular focus on forward-looking ideas concerning innovative or as yet undeveloped technologies. Whilst there is an expectation that authors will be well grounded in philosophy, law or political science, consideration will be given to future-orientated works that cross these disciplinary boundaries. The interdisciplinary nature of the series editorial team offers the best possible examination of works that address the 'ethical, legal and social' implications of emerging technologies.

For more information about this series, please visit: https://www.routledge.com/Emerging-Technologies-Ethics-and-International-Affairs/book-series/ASHSER-1408

Contemporary Technologies and the Morality of Warfare
The War of the Machines
Jean-François Caron

American Security and the Global War on Terror
Edwin Daniel Jacob

The Politics of Technology in Latin America. Volume 1
Data Protection, Homeland Security and the Labor Market
Edited by Avery Plaw, Barbara Carvalho Gurgel and David Ramírez Plascencia

The Politics of Technology in Latin America. Volume 2
Digital Media, Daily Life and Public Engagement
Edited by David Ramírez Plascencia, Barbara Carvalho Gurgel and Avery Plaw

The Disruptive Impact of Lethal Autonomous Weapons Systems Diffusion
Modern Melians and the Dawn of Robotic Warriors
Austin Wyatt

The Disruptive Impact of Lethal Autonomous Weapons Systems Diffusion

Modern Melians and the Dawn of Robotic Warriors

Austin Wyatt

Routledge
Taylor & Francis Group

LONDON AND NEW YORK

First published 2022
by Routledge
2 Park Square, Milton Park, Abingdon, Oxon OX14 4RN

and by Routledge
605 Third Avenue, New York, NY 10158

Routledge is an imprint of the Taylor & Francis Group, an informa business

© 2022 Austin Wyatt

British Library Cataloguing-in-Publication Data
A catalogue record for this book is available from the British Library

Library of Congress Cataloging-in-Publication Data
A catalog record has been requested for this book

ISBN: 9781032001531 (hbk)
ISBN: 9781032001555 (pbk)
ISBN: 9781003172987 (ebk)

DOI: 10.4324/9781003172987

Typeset in Times New Roman
by Deanta Global Publishing Services, Chennai, India

This book is dedicated to the memory of Professor Donald Iverson. Don was an invaluable mentor during my time as an undergraduate. While he did not get to see this book published, I owe a great debt to his wisdom and will continue to strive to make the most of this pathway that he helped me pursue.

Contents

Figures

Tables

Glossary of commonly used abbreviations

ACT	Adoption Capacity Theory
ADMM	ASEAN Defence Ministers' Meeting
AMT	Autonomous Military Technology
ARF	ASEAN Regional Forum
ASEAN	Association of Southeast Asian Nations
AWS	Autonomous Weapon System(s)
CCW	Convention on Certain Conventional Weapons
COTS	Commercial Off the Shelf
ICRC	International Committee of the Red Cross
IHL	International Humanitarian Law
IHRL	International Human Rights Law
LAWS	Lethal Autonomous Weapon System(s)
NSAG	Non-State Armed Groups
RMA	Revolution in Military Affairs
RPA	Remote Piloted Aircraft
SAF	Singapore Armed Forces
TNI	Tentara Nasional Indonesia (Indonesian Military)
TNI-AD	Indonesian Army
TNI-AL	Indonesian Navy
TNI-AU	Indonesian Air Force
UAV	Unmanned Aerial Vehicle(s)
UCAV	Unmanned Combat Aerial Vehicle(s)
UCV	Unmanned Combat Vehicle(s)
UGV	Unmanned Ground Vehicle(s)
UMV	Unmanned Maritime Vehicle(s)
UUV	Unmanned Underwater Vehicle(s)

1   Introduction

There are but two powers in the world, the sword and the mind. In the long run, the sword is always beaten by the mind – Napoleon Bonaparte

Built into a concrete outcrop in one of the most dangerous places on Earth, an electronic eye stares out from its squat housing, the long nose of its heavy machine gun deterring potential infiltrators in the absence of the camouflaged conscripts that previously patrolled this hazardous stretch of the De-Militarised Zone (Parkins, 2015).[1]

While remote-operated and human-supervised weapons had been previously deployed, the Super Aegis II was among the first examples of a weapon platform with the capacity to exercise effectively independent control over the operational selection, identification and engagement of human targets.

Removing the human from the decision to employ lethal force raises a number of serious moral, ethical and legal questions that continue to dominate the debate at the semi-regular meetings of the United Nations-sponsored Group of Governmental Experts on Lethal Autonomous Weapon Systems (LAWS). Setting aside the remaining technological barriers (Anderson, 2016), multiple states have declared their position on LAWS (and militarised applications of Artificial Intelligence [AI] more broadly). The problem with the formal discussions to date is that, while there have been some notable contributions by smaller states to the formal discussions, it has remained focused on great power perspectives.

The overarching purpose of this book is to address that gap in the international discourse by presenting the first full-length examination of the role Asia-Pacific middle powers will have on the development and proliferation of autonomous weapon systems, as well as an examination of how this innovation will shape the role of middle powers in the growing great power competition between the United States and China. This book, therefore, takes a step beyond the current public and scholarly debate to focus on how the emergence of LAWS will affect the regional security of South-East Asia.

This is certainly not the first time that the development of a particularly disruptive military technology has challenged the paradigm of conflict and international power. By undermining the existing power projection paradigm, such innovations

DOI: 10.4324/9781003172987-1

can enable emerging states to challenge a dominant hegemon (Vickers and Martinage, 2004). When this occurs at the major state level, the resulting competition between a challenger state and the dominant hegemon results in a transition of hegemonic power. This transition increases tension, and often sparks conflict, between the rising power and the dominant hegemon (Allison, 2017). Historically high economic, technological or knowledge-based barriers (Horowitz, 2010), such as sophisticated composite materials or specialised knowledge, have injected a level of structural stability into this process by constraining the hegemonic conflict from spreading to minor powers. However, the dual-use nature of the enabling technologies (such as machine learning coding, computer processing power, relevant datasets and mass-produced sensors) distinguishes LAWS from previous major military innovations. While reliable, fully autonomous lethal weapon systems remain beyond the capabilities of modern technology, other forms of autonomous and semi-autonomous military technologies have much lower technological requirements. This book, therefore, considers the full range of autonomous military technology, recognising the potential impact of derivative or copied autonomous weapon systems and the diffusion of the underlying technology to other actors. The core research aim of this book is to explore how LAWS, with their lowered barriers to initial proliferation, to multiple middle powers could impact the transition of hegemonic power within the geographic confines of Southeast Asia.

Although the discussion around LAWS is far more advanced in 2021 than it was when the Super Aegis II was unveiled six years ago, the parameters of that discussion have not sufficiently shifted and remain focused on great powers and international law. This has left a gap in our understanding of the impact of autonomous weapon system proliferation when it comes to the actions and perspectives of small-middle power states. This gap is particularly damning in the case of Southeast Asia, which is a region of growing global economic and geopolitical importance that straddles some of the hottest potential global flashpoints and key corridors of international trade. Despite this importance, the Campaign to Stop Killer Robots only opened its Southeast Asian satellite arm in July 2019 (Picard, 2019), and only three Association of South-East Asian Nations (ASEAN) member states participated in the meeting of the Group of Governmental Experts on Emerging Technologies in the Area of Lethal Autonomous Weapon Systems in the following month. Furthermore, as of September 2019, no ASEAN member state has released a statement codifying their position on the merits of a ban aside from a short statement in 2018 issued by the Non-Aligned Movement. Far from staying idle during this period, however, Indonesia and Singapore (as leading ASEAN member states) have taken active and overt steps that indicate a clear desire to integrate increasingly autonomous and remote-operated weapon systems into their ongoing military modernisation efforts.

Therefore, while the emergence of increasingly autonomous military technology will change our understanding of warfare and power projection, the question remains as to how their rapid proliferation will affect the balance of power in South-East Asia, especially once the smaller states gain access to autonomous weapon systems. This book responds to this gap, analysing and exploring how the Singaporean and Indonesian response to the emergence of increasingly autonomous weapon systems will impact relations of power at the regional level, and how this will shift their role in Sino-American hegemonic competition.

1.1 Research questions

Responding to this core research puzzle, this book centres on a primary research question:

1. What impact will the adoption of autonomous military technology by the middle power states within South-East Asia have on regional security in the Asia-Pacific?

Three secondary questions are derived from the primary question:

A. How did the key Southeast Asian states respond to the proliferation of remote-operated Unmanned Combat Vehicles (UCVs), and how is this influencing their approach to increasingly autonomous weapon systems?

Much of the military innovation and diffusion literature, as well as the military history literature, contends that the demonstration of a new major military innovation presents states with a stark choice; to adopt the new innovation or to allow the first mover state to gain a power projection advantage. As Horowitz (2010) indicates, the French navy of the late 19th century was the first mover in numerous major innovations in naval warfare, including the submarine, and yet the British cemented their status as the premier naval power. Where an early mover state has failed to capitalise on an emergent major military innovation, it has historically allowed a rival state to master the innovation for its own benefit (Silverstein, 2013). Furthermore, military policymakers have historically demonstrated a tendency to view the emergence of a major military innovation with reference to similar precursor innovations. Therefore, any analysis of an emerging military innovation should begin by outlining the development, and diffusion of its precursor innovation. In the case of LAWS, the precursor innovation is UCVs, which are distinguished by the fact that their 'critical functions' remain under the control of a human operator, albeit remotely.

B. What factors will influence the rate of LAWS proliferation into South-East Asia?

This question delves deeper into the puzzle, comparing how the lower barriers to proliferation will affect South-East Asian state response to a future LAWS demonstration point. There are a number of theories of military innovation and diffusion; Grissom (2006) presents a concise summary of the leading theories. This book adopts a neo-realist perspective of state behaviour and utilises elements from two leading theories of military innovation: organisation theory (Goldman and Andres, 1999) and adoption-capacity theory (Horowitz, 2010). These theories emphasise the importance of financial and organisational capacity barriers to determine which states are likely to adopt a particular innovation. When a major military innovation requires a high level of resources or is reliant upon controllable components, it is unlikely to be adopted by smaller states in the short term. Aircraft carriers, stealth aircraft and intercontinental ballistic missiles are all examples of such innovations. Equally, innovations that rely on specialised knowledge or require major doctrinal changes to be successfully deployed also diffuse slower (Horowitz, 2010). Finally, domestic pressures or a cultural aversion to the innovation, within the military or across the society, also reduces the likelihood that a given innovation will be adopted (Goldman and Andres, 1999). An example of this effect is nuclear weapons, which require specialised knowledge and skills, have high resource requirements and are the subject of strong cultural aversion. In contrast, cyber warfare has far lower barriers in each category and therefore proliferated at a significant speed. This book theorises that autonomous weapon systems, which have low barriers to proliferation, will follow a slower but similar proliferation pattern to cyberwarfare or remote-operated military aircraft. This would be the first time that an innovation as potentially disruptive as LAWS proliferated at such a fast rate, potentially de-stabilising any hegemonic conflict that autonomous military technology enables.

C. How are the expected capabilities of LAWS influencing the South-East Asian security environment?

The objective of this question is to engage critically with the first aspect of the difference between autonomous military technology and prior major military innovations, proliferation to smaller, middle power states in its early stages. While major previous innovations have enabled transitions of hegemonic power, sometimes involving conflict, their high barriers to proliferation have ensured that only the major powers have been able to acquire major military innovations in their early stages of development or proliferation. Prior to the development and proliferation of autonomous weapon systems, therefore, major powers were able to exert influence over smaller powers by virtue of a dominant position in the development and supply of related weapon systems.

A neo-realist approach holds that at the core of the international system are power relationships, an intricate web of economic, security and cultural ties that allow for day-to-day international relations. Relative international stability is derived from the maintenance of a balance of power between the various states, supported by a network of norm-based agreements (Mearsheimer, 2013). This balance is maintained by a collection of major powers, of which one is considered the overall hegemon. In the post-Cold War balance of power, the hegemon is the United States. Central to the theory of hegemonic power is the concept that the hegemonic state can only maintain its status by being the strongest and wealthiest state, a requirement that inevitably results in tension between the existing hegemon, which is clinging to its position of power, and the emerging power, which feels constrained by its rival (Allison, 2017).

A key trigger of conflict between major powers, and a common determinant of their outcome, is the adoption of paradigm-shifting military innovations. A shift in the paradigm of warfare and power projection creates an opportunity for a challenger state to disrupt the international system, by using dominance over the emerging technology or capability to compensate for a comparative lack of conventional military power. For example, instead of expending exorbitant resources in developing an aircraft carrier fleet, a challenger could develop autonomous surface vehicles to swarm and disable carrier battle groups in shallower waters. Alternatively, a smaller state could cripple a rival's air force by sending a swarm of small, cheap, autonomous aircraft to interfere with enemy aircraft as they take off (Gaub, 2011). Historically, the increase in a state's ability to project power, as well as their prestige, has disrupted the dominant balance of power, enabling the rising powers to challenge the existing ones (Gilpin, 1988). Although earlier adoption does not guarantee ascendance for the challenger (Silverstein, 2013), the state that proves superior at integrating the major military innovation has historically gained comparative influence in the post-conflict balance of power.

The end state of hegemonic conflict is the establishment of a new balance of power. Historically, this has been a stable, albeit occasionally violent, process. This is because the initial diffusion and adoption of prior major military innovations were limited to large states, ensuring their comparative advantage over minor states and limiting the scope of the conflict. Historically, competing major powers have instead bound minor powers to their cause, building supporting coalitions within their claimed sphere of influence. While the ideal situation for a middle power is to exist in a stable dual hegemony, they rarely have the security or economic capacity to alienate a potential hegemon (Ikenberry, 2016). This effect, while disenfranchising to the minor states, reduced their security dilemma because their allegiance to a major power provided protection and stability. Among the earliest examples of this effect was the Melian Dialogue, while a more recent example was the consolidation of alliances during the Cold War. This is currently the case in the Asia-Pacific, where the ASEAN member states are finding it increasingly difficult to maintain a split allegiance, relying on the Chinese economic influence and the US security partnership. In the event that major states were not able to maintain their comparative advantage in terms of an emerging military innovation, there is no real scholarly consensus on how minor regional powers would react, although the rapid proliferation of remote-operated drones would suggest an intensification of the existing regional arms race unfolding in Southeast Asia. This research question focuses on exploring the potential responses of middle and minor powers in the region with the goal of promoting more active engagement with this issue by policymakers and fellow scholars.

1.2 Hypothesis

The core hypothesis of this book is that the uniquely low diffusion barriers of autonomous military technology comparative to previous major military innovations will allow for rapid proliferation to middle power states in Southeast Asia. The rapid diffusion of a disruptive military innovation to non-major powers, which has never occurred before, will de-stabilise the emerging hegemonic conflict between China and the United States, leading to an unstable balance of power in the region. The presence of such a disruptive weapon system in the armouries of minor, rising regional powers will raise the security dilemma of neighbouring states and make multiple regional conflicts more likely. This will contribute to increased intra-regional conflict and instability in a region of vital geopolitical importance to the global economy and security.

Despite their revolutionary nature, previous major military innovations have, for a number of reasons, reinforced an underlying structural paradigm within international power relations by excluding minor powers from active participation in major transitions in the balance of power. Instead, minor powers were subsumed to the will of a major power for the duration of the hegemonic conflict. The Melian Dialogue has become illustrative of this effect, lacking the resources and technological superiority required to play an independent role they are subjected to the will of one side or the other for the duration of the conflict. As a more modern example, consider the broadly bi-polar consolidation of states during the Cold War. During the Cold War, the Non-Aligned Movement emerged to counterbalance the influence of the competing hegemonies over smaller states in the global south. Joining a balancing coalition in this manner allows weaker states to offset the influence of the more individually powerful competing hegemonies (Hamilton and Rathbun, 2013). Regional supranational trade/security organisations such as ASEAN and the African Union are the modern successors of this Non-Aligned Movement.

Breaking down this core hypothesis, (A) Southeast Asian states are expected to have responded to the proliferation of remote-operated vehicles as secondary adopters, sitting within the early majority section of an S-curve. However, it is expected that Indonesia and Singapore would attempt to integrate these systems in emulation of their larger peers once the underlying technology had matured and adoption barriers had sufficiently fallen. Based on the examples of prior military diffusion, it is hypothesised that the initial perceptions of LAWS by both the Indonesian military (TNI) and Singaporean Armed Forces (SAF) TNI and SAF would be heavily influenced by their existing platforms and developing experience with remote-operated platforms. Secondly (B), it is hypothesised that neither case study state will have the comparable resource capacity to that of the United States or China; however, it is expected that organisational barriers will be a more significant challenge for ASEAN militaries. Finally (C), it is anticipated that the early diffusion of autonomous military technology will enable early-adopting middle power members of ASEAN to retain a greater level of independent action, avoiding the Melian's dilemma in an emerging hegemonic conflict. Unfortunately, this would then increase the security dilemma of middle and minor powers in the region. Defensive neo-realism indicates that these states would then attempt to increase their power to secure their influence in the region (Mearsheimer, 2013). Both options would complicate a hegemonic conflict and increase regional instability.

1.3 Methodology and research design

The core of the research design of this book is a case-study-based approach, supported by process-tracing. The core case studies for this book are Indonesia and Singapore, the leading regional middle powers and ASEAN member states. A case-study-based methodology has been favourably reviewed in multiple meta-analyses as particularly suited for studying military diffusion and has been utilised in the fields of policy diffusion, military innovation and disruptive commercial innovation.

In order to ensure analytical validity in its projective analysis, this book limits itself to modern, publicly accessible and sceptically examined data, sourced through a combination of analyses from defence research bodies, civilian state agencies, non-government think tanks alongside traditional academic literature. This information is analysed through a composite theoretical framework that incorporates elements of Revolution in Military Affairs, Adoption-Capacity Theory, Organisational Innovation, Precursor Wars and the Thucydides Trap. While a novel construction, this theoretical framework remains grounded by its neo-realist security studies theoretical roots.

The book structure directly reflects its theoretical framework, and its chapter progression reflects the broad progression of disruptive military innovations through the four stages that comprise the composite framework. In addition to answering each research sub-question in turn, using these four stages as an analytical skeleton ensures a logical progression of the book and delineate between the analytical and conceptual components.

The first phase in this book's theoretical framework is Foreshock: it covers the development of precursor technologies (which may in their own right be initially lauded as significant innovations), their impact on the development of a disruptive weapon innovation and their proliferation once the precursor becomes normalised. This stage is addressed in Chapter 4, focusing on the response of Indonesia and Singapore to the proliferation of remote-operated UCVs. The second phase is Innovation; it engages with the initial development of the revolutionary technology and the emergence of new strategic or operational doctrine that capitalises on the invention, leading to the achievement of operational praxis. This stage is reflected in Chapter 5, which details the key actors in the development of the hardware and software components of this innovation, as well as evaluating the current progression towards a demonstration point. The third stage, Adoption, begins with the demonstration point of the innovation, which triggers states to respond to the shift in the balance of power. This relates directly to the core research questions of this book and, as such, is the main application of the two case study states in

Chapters 6 and 7. The final stage is Impact, which covers how the international community goes about the ongoing development of the initial disruptive innovation, the regional instability caused by its diffusion and the possibility of a transition of hegemonic power, at least on a regional level. This stage is reflected in Chapter 9 of the book, which evaluates the impact of LAWS proliferation in Southeast Asia and the hegemonic transition conflict between the United States and China.

1.4 Contribution

This book is situated at the intersection of three key theoretical fields: diffusion of innovation, revolution in military affairs and international power transition. Even though the study of LAWS has received the attention of eminent scholars in recent years, this has largely remained focused on great powers, international law and ethical questions. Therefore, a gap remains in understanding LAWS proliferation from the perspective of Southeast Asian states, as well as how this proliferation would influence the post-demonstration point period of a concurrent hegemonic conflict.

Overall, this book makes three major contributions to the existing scholarly literature, which are intended to also support the policymakers in ASEAN states during the current incubation period. While these contributions are focused on autonomous military technology and Southeast Asian actors, this research is more broadly applicable to improving scholarly understanding of how major military innovations with low adoption barriers proliferate within a complex regional environment.

In prior scholarly works on hegemonic conflict, smaller states have been typically relegated to a minor role, subsumed by the goals of the major states. This omission is informed by the offensive neo-realist theoretical framework for power transition theory as well as prior transitions of hegemonic power, which are presented as largely binary confrontations between an existing hegemon and an emerging rival. Without the capacity to generate sufficient conventional power to assert their own interests over those of major states and without the ability to subvert this equation with a major military innovation, the minor states are forced to either bandwagon or accept direction from one of the competing hegemons. Within the emerging body of literature that examines the proliferation of unmanned combat aircraft, there is also a clear focus on major powers. Some leading scholars have dismissed the potential impact of unmanned military systems (remote-operated or autonomous) on the basis that ASEAN member states lack the data infrastructure to emulate the United States' use of unmanned aircraft, neglecting the very real potential impact of diffusion of the underlying technology or proliferation of derivative weapon systems. Interestingly, while the process leading to hegemonic conflict is called the 'Thucydides trap', prior scholarly works have not considered how autonomous military technology could affect the modern Melians.

Therefore, the first main contribution of this book lies in bridging these gaps in the existing literature. This book evaluates how the diffusion of autonomous weapon systems will impact security in Southeast Asia from the under-researched perspective of middle power states. The stability of Southeast Asia is of key geo-political importance to three of the seven states that are openly developing LAWS as well as being a focal point of nuclear tension. Crucially, the United States and China (which are both rapidly developing LAWS) are deeply invested in Southeast Asia. While the emerging hegemonic conflict between China and the United States is most apparent on the Korean Peninsula, which is a key geopolitical flashpoint for the competing hegemonic powers, this book looks to the nearest concentration of rising middle power states, ASEAN. Southeast Asia, with its rapidly growing, mutually suspicious states and the multitude of violent non-state groups, is a more suitable region to focus on based on this book's underlying puzzle.

The second key contribution of this book is increasing the scholarly under-standing of the socio-political and cultural influences that impact the proliferation and adoption of major military innovations, such as LAWS. Previous historical major military innovations demonstrate that merely possessing superior technol-ogy is insufficient for a state to maintain its power during the emergence phase. The states' that are most successful with disruptive military innovations have historically been the states that are best able to match technological advance-ment with sufficient operational flexibility and domestic engagement to modify their existing strategic doctrine to capitalise on the entire innovation. Departing from the neo-realist perspective, this book contributes a greater understanding of the cultural, organisational and political influences on a state or non-state actor's decision to adopt the major military innovations as well as how these factors will specifically affect the adoption of LAWS in Southeast Asia.

While scholars may argue about the precise extent of its influence, technology has undeniably played a major role in human conflict. Even the earliest tribal humans constructed tools to aid them, whether as a hunter or a warrior. Over time, these tools have become more sophisticated. Although improving military technology has reduced reliance on soldiers and improved the ability of a given state to project power, this has often come at the cost of investment in expensive, highly advanced military systems, such as aircraft carriers or stealth aircraft, con-centrating power in a handful of wealthy states. While these military innovations eventually diffuse and proliferate, the high financial and organisational barriers to adoption slows this process, maintaining their power differential (Horowitz, 2010). The low entry barriers to increasingly autonomous unmanned platforms would circumvent this slow, stable process, a factor that has not been deeply explored in prior scholarly works. This book provides a different perspective on major military innovation and its impact within a globalised world, and in the context of modern conflict, this approach encourages further exploration of other major innovations that share a similar technological base, such as cyber warfare.

1.5 Conclusion

The development of increasingly autonomous military technologies is already well underway and yet the international community remains locked in a debate that has largely overlooked the participation of middle power and Southeast Asian states. This is despite the fact that the lower diffusion barriers of unmanned and increasingly autonomous systems create a unique opening for meaningful, albeit limited, secondary adoption rising middle power states. This book, therefore, takes a step beyond the current public and scholarly debate to focus on how the emergence of LAWS will affect the regional security of South-East Asia, which is of vital geopolitical and economic importance. To understand the impact of LAWS it is vital to understand the depth of the rapidly expanding scholarly litera-ture that focuses on the ethical, technological, legal, practical and moral impacts of this technology.

Note

1 The author would like to thank Professor Mark Chou, Professor Michael Ondaatje, Dr Amanda Alexander, Professor Kenneth Anderson, Professor Michael Horowitz and Dr Jai Galliott for their feedback and support during the preparation of this book. I would also like to thank the anonymous reviewers and my copy-editor. All mistakes remain solely the fault of the author.

2 Literature review and theoretical framework

Hope is not a strategy – Julie Bishop, Former Australian Minister for Foreign Affairs

2.1 Introduction

The development of lethal autonomous weapon systems (LAWS) will have a profound impact on the conduct of warfare. However, there is a tendency in the public discourse, particularly in the West, to regard revolutionary advancements in technology as somehow appearing out of the blue. The reality is that even the most paradigm-shifting innovations are shaped by their context and predecessors. Understanding the impact of these influences on decision-makers, as well as technology development, is an often overlooked, but important step in any innovation analysis. Therefore, any analysis of autonomous military technology must go beyond the specifically applicable literature to engage with scholarly works that more broadly examine major military innovations and how these innovations affect relations of state power.

The purpose of this chapter is to highlight some of these key influences and explore how the existing literature has engaged with the questions posed by each. Seemingly counter-intuitive for a book focused on technological advancement, this chapter begins with an exploration of the history of remote-operated weapon technology. The current public debate surrounding lethal autonomous weapon systems tends to disregard the fact that humans have a long and very relevant history of utilising remote-operated systems both on and off the battlefield. This section demonstrates that the development of LAWS is not occurring in a vacuum. It shows how historical cases of remote-operated weapons have affected the course of military history as well as how these prior forays are impacting the development of fully autonomous weapon systems.

Equipped with a firm historical grounding, this chapter proceeds into a comprehensive review of core questions, themes and debates that have come to dominate the comparatively young, study of lethal autonomous weapon systems. The main three themes addressed in this section are the question of how to define lethal autonomous weapon systems in a manner that is sufficiently specific to

DOI: 10.4324/9781003172987-2

allow for effective regulation and that would be supported (or at least respected) by the majority of states; the ongoing debate regarding whether lethal autonomous weapon systems inherently offend the principles of international humanitarian law, how to regulate their development and deployment in future conflicts and the merits of the international community imposing a pre-emptive ban.

The final section of this chapter takes a step back from the specific technology at the centre of this book in order to provide a thorough summary of prior works in the fields of military innovation, theories of technology diffusion and the role of major military innovations in power transition and regional stability. These themes form the three pillars of the theoretical framework that support this book and directly informed the development of the Disruptive Weapon Innovation framework discussed in the next chapter.

2.2 Historical overview of the development of autonomy in weapon systems

The history of autonomous weapon systems (AWS) begins with the pursuit of more effective and efficient methods of inflicting violence. In military history, there is a clear pattern of innovations that increase the emotional and physical distance between combatants (Grossman and Christensen, 2007). Historically, this has primarily involved increasing the range and lethality of weapons. This pursuit has fascinated military planners, soldiers and civilian leaders for centuries, driving the development of firearms, artillery and, of course, military robotics. Although the scale of the remote operation was initially limited by technological progress to relatively rudimentary radio control, wire control and/or teleoperation, the rise of satellite technology, increasingly powerful computing technology and advancements in robotics have propelled the human combatant further from the immediate battlefield. An effective analysis of the impact of autonomous weapon systems must therefore start by examining the historical evolution of remote-operated weapons.

Although there is an immense amount of scholarship available regarding human nature, it is safe to assert that cultural constructs play an important role in innovation. The development of unmanned systems is no exception. Although the term 'robot' was coined in 1920 (Hockstein, Gourin, Faust and Terris, 2007), the underlying concept can be traced to Greek mythology. Among the best-known examples are the automatons, which were created by Hephaestus or Daedalus (depending on the source). In a remarkably similar manner to modern robotics, these automatons were said to be capable of independently performing a given task or function, what we now call task-based autonomy. Examples of automatons range from Talos, the golden giant who patrolled Crete and threw boulders to defend its island from pirates to Khryseos and Argyreos, which were 'deathless forever and unageing' guard dogs created by Hephaestus from gold and silver. Greek mythology also makes reference to less warlike automatons such as the Tripodes Khryseoi, a set of 20 golden tripods that independently served food at the Olympian feasts (Atsma, 2000). Partly inspired by these myths, a number of non-electro-mechanical automatons were designed over the early modern period. However, these were mostly individual artisan pieces without widespread military application.

The first weaponised use of remote-operated systems occurred in 1849. Faced with a terrain that was unfavourable to direct bombardment, the Austrian army built a fleet of 200 unmanned paper balloons that each carried approximately 15 kg of explosives. The goal was to use trailing copper wires to remotely release the explosives en-masse over Venice. While some bombs were successfully delivered, an errant wind sent many off course and even blew several back into the Austrian lines. While not totally successful, this is considered the first example of unmanned aerial bombardment. Although both sides made extensive use of balloons during the American Civil War for surveillance, there were limited instances of balloons being used remotely for bombardment, and their advocates were largely dismissed as eccentrics. The first remote-controlled torpedo was demonstrated in 1866, sparking further development of remote-operated munitions and boats. One of the more widely deployed was the Brennan torpedo, designed by an Australian in the late 1870s (Everett and Toscano, 2015).

The advent of radio control was a major step in the development of unmanned military technology. In the late 1890s, Nikolas Tesla demonstrated the capacity of remote-controlling vehicles by radio signal. With his trademark showmanship, Tesla used radio control to make a large iron ship follow shouted directions from the crowd (Finn and Scheding, 2012). Tesla's stated goal was to militarise this advancement, specifically for use in the Spanish–American War. Tesla had come to the belief that *Telautomats* (remotely controlled robots) were the future of conflict and, in a remarkably similar leap of logic to Richard Gatling, believed that they would make the conflict so terrible that states would no longer start wars (Chivers, 2010).

It was not until the World Wars that the development of remote-operated systems is generally held to have begun in earnest. The First World War saw the introduction of some key concepts, but unmanned systems had a minor impact. In a revival of the fireboat concept, the German navy made limited use of remotely operated boats packed with explosives (Williamson and Palmer, 2012). On the Allied side, the Kettering Bug and Wickersham Land Torpedo were the respective spiritual forbearers of cruise missiles and unmanned ground vehicles (Finn and Scheding, 2012). During the interwar period, the Japanese army developed the Nagayama remote control tank and the Type 98 Mini Engineer Vehicle 'Ya-I Go', a remote-operated engineer vehicle that bore a striking resemblance to the modern PackBot. Ultimately, however, neither design saw combat (Krishnan, 2016).

The Second World War saw the first use of a remote-operated unmanned ground combat vehicle (UGCV) in frontline combat, the Russian TT-26 Teletank. Designed in the 1930s, the Teletanks were T-26 light tanks that had been modified for remote radio control and were armed with flamethrowers and heavy machine guns. They were generally remotely operated from up to 1,500 m away by a crew that rode in a second T-26 (Finn and Scheding, 2012). The Red Army had two battalions of Teletanks that saw action in the Winter War and the opening stages of

Operation Barbarossa. While one of the battalions was destroyed by German aerial bombardment in early 1940, the other participated in the defence of Moscow before being converted back for human operation. Due to the sophistication of their control system, the TT-26s were considered highly classified and crews were ordered to fire on Teletanks that were in danger of capture (Turner, 2016). Another remote-operated weapon from this period was the German Goliath tracked mine, which was based on a captured French prototype. The Goliath was effectively a bomb on tracks that were operated remotely through a control wire. It was not well received due to its slow speed and vulnerability to small arms fire. The US Army and Navy made the first mass purchase of an unmanned aircraft during Second World War, purchasing 15,000 Radioplane OQ-2 to be used as practice targets (Finn and Scheding, 2012). This became the primary role for unmanned aircraft in the Cold War period.

During the Cold War, the development of unmanned systems slowed substantially, largely due to lack of military interest. Despite this, the period did include the conversion of the Firebee target aircraft into the first military unmanned surveillance aircraft, the Lightning Bug, in response to the 1962 U2 shootdown over Cuba (Newcome, 2004), and its subsequent deployment in the Vietnam War (Keane and Carr, 2013). The Lightning Bug platform was utilised extensively, flying 3,435 missions between August 1964 and June 1975, reaching an operational tempo of four missions per day during Operation Linebacker (Newcome, 2004). The Lightning Bugs, in variously modified forms, performed a wide range of roles over this period, including collecting photographic and signals intelligence, dropping leaflets and engaging ground targets with Maverick missiles. As with modern Unmanned Aerial Vehicles (UAVs), the survivability of Lightning Bugs in effectively contested airspace was significantly lower than would be acceptable in manned platforms, especially after Russian-made Surface-to-Air-Missiles (SAMs) were installed in North Vietnam, with over half of the 1,016 platforms destroyed (Newcome, 2004). Despite the success of its deployment, this early unmanned aircraft project ran afoul of the ingrained organisational culture within the US Air Force, which focused influence on pilots (who were further separated between the tactical and strategic air commands). Despite a number of efforts to ingratiate pilots to the systems, including referring to the UAVs as remotely piloted vehicles (which is now the US Department of Defense's [DoD] preferred term), this project was sacrificed to peacetime spending cuts in favour of manned platforms.

The Vietnam War also included the first deployment of a remote-operated rotary-wing uninhabited aircraft, the QH-50 DASH. The DASH was originally designed as a range-extender and force-multiplier for destroyers in their Anti-Submarine Warfare (ASW) role. The DASH was intended to be launched from a destroyer that detected an enemy submarine beyond the effective range of its onboard anti-submarine armament. To this end, the DASH was equipped with two Mark 44 homing torpedoes that would be launched by a remote pilot in consultation with sonar operators on the destroyer. It is noteworthy that subsequent upgrade projects included the addition of low-light and all-weather sensor packages, and the installation of ground-attack munitions (including a grenade launcher and guided rockets). The DASH was used by the US Navy in Vietnam primarily as a fire control platform, providing guidance and alterations to naval gunners firing in support of ground troops (Newcome, 2004). The Vietnam War saw a number of important firsts for the use of remote aircraft by the United States, but it was also a clear example of an innovation being abandoned to peace-time cuts due largely to internal organisational politics.

The Israeli use of unmanned aircraft in the Yom Kippur War and the 1982 Lebanon War were the key milestones in UAV development. In the former case, the Israeli Air Force military countered the strong Syrian air defences in the Golan Heights by tricking them into exposing their positions and wasting ammunition with a wave of unmanned target aircraft (Doyle, 2016). In the 1982 Lebanon War, *Scout* UAVs were instrumental in the first defeat of an air defence system based on the Soviet SAM batteries by a Western air force (Libel and Boulter, 2015); demonstrating to the world the military value of unmanned aircraft beyond surveillance. The value of pairing real-time surveillance with a capability to agilely direct force appealed to the Israeli military and intelligence services tasked with responding to the ongoing attacks by Palestinian organisations. A decade after the Lebanon War, the Israeli Air Force (by this time to primary operator) and Israeli Aerospace Industries (still the main manufacturer of UAVs in Israel) had equipped their UAVs with laser designators for coordinating strikes by manned aircraft and were training operators to mark rapidly moving targets (Libel and Boulter, 2015). In the same year, foreshadowing the US Air Force's decision almost ten years later, Israeli authorities utilised a UAV to coordinate the targeted killing of Abbas Al-Musawi by an Apache helicopter (Libel and Boulter, 2015).

The development of modern remote-piloted military aircraft began in the 1980s with the use of AAI RQ-2 Pioneers in the First Gulf War (Polmar, 2013). The Pioneer UAVs were primarily used as spotter planes to direct naval gunfire, which led to the first instance of combatants attempting to surrender to an unmanned aircraft (Newcome, 2004). This was followed closely by the development of the, now well-known, MQ-1 Predator by General Atomics in partnership with Big Safari, the US Air Force's special weapons innovation programme.[1] First used in Kosovo as a surveillance tool, the decision to arm these unmanned aircraft was taken in June 2000. With the first lethal missile strike from an MQ-1 Predator on 7 October 2001 (Whittle, 2014), the development of the Unmanned Combat Air Vehicle (UCAV) had come to fulfilment. This was arguably the demonstration point of the UCAV and almost immediately triggered a race among other powers to develop their own UCAVs. At the time of writing, almost 20 years after the first strike, the US has retired and replaced the Predator, but UCAV technology has spread to over 80 states (Sayler, 2015) and there is a rapidly expanding market for the civilian commercial remotely piloted airport that has been drawn on by state and non-state actors.

The historical development of remote and unmanned systems by militaries is often minimised in discussions of the potential deployment of LAWS. Yet this historical background provides crucial lessons that help contextualise potential reactions by civilian and military policymakers towards unmanned, autonomous weapons as well as insights into how their development may proceed. The final historical example referred to above, the UCAV, is of direct relevance to this book. If autonomous weapon systems follow a similar diffusion and proliferation pattern to that followed by UCAVs over the past decade, the impact would be severe. On that note, it is useful to turn to a comprehensive review of the expanding body of scholarly and technical literature available that examines lethal autonomous weapon systems themselves.

2.3 Distinguishing lethal autonomous weapon systems

When exploring the impact of an innovation, it is vital to understand the characteristics of that innovation and, perhaps more importantly, how to differentiate the innovation from similar products. This is especially important when trying to understand the impact of a disruptive innovation (in the civilian realm) or a Revolution in Military Affairs (RMA) (in the military space). Clear definitional boundaries are also vital for policymakers, who are tasked with developing regulatory responses, and businesses, who are trying to capitalise on the resulting market shift. Setting definitional limits or criteria on emerging technologies also has a potential political impact, with stakeholders aiming to influence acceptable definitional elements to shape future norms, laws, discourse and state action. As an example, consider that there is still no universal definition of 'terrorism' despite immense funding devoted to terrorism research over the almost two decades since 9/11. It is therefore unsurprising that no universally agreed definition has emerged for lethal autonomous weapon systems.

Instead of entering this ongoing debate, this book synthesises elements drawn from a selection of prominent definitions to form a working definition of autonomous weapon systems for the purposes of analysis.

> A fully autonomous Lethal Autonomous Weapon System (LAWS) is a weapon delivery platform that is able to independently analyse its environment and make an active decision whether to fire without human supervision or guidance. (Wyatt and Galliott, 2018)

Whether a given definition would be considered 'prominent' in this respect is largely dependent on the extent to which it was cited in the scholarly literature, whether it was referred to in the official statements issued after each meeting of the Group of Governmental Experts on LAWS, and the extent of the author's broader contribution to military diffusion studies or AWS research. I also draw on the extant definitional analyses published by authors including Conn (2016), Jenks (2016) and Horowitz (2016).

Regardless of the specific definition, it is important to note at the outset that it is not realistic to consider autonomy in the robotics field in binary terms, rather it is much more analytically effective to consider autonomy as a function-based spectrum where human interaction remains present at some point (even if it is limited to the production or strategic deployment stages) (Anderson, 2016). This book builds on this basic understanding and, building on a division first used by Human Rights Watch, uses a three-category system for grouping weapon systems based on the level to which certain 'critical functions' are handled without meaningful human input. Before moving on to describing these categories, it is important to define the RMA itself, lethal autonomous weapon systems.

2.3.1 The ongoing debate on how to define a lethal autonomous weapon system

Developing a definition for a complete lethal autonomous weapon system is arguably one of the major stumbling blocks to developing an effective international response to the emergence of increasingly autonomous military technology, whether this is regulation or a developmental ban. As a result of political and practical issues, the international group of experts convened by the United Nations has been unable to generate a definition of autonomous weapon systems that would be universally agreed upon or that could operate as the basis for a pre-emptive development ban. In this gap, various actors, from states to arms companies to scholars, have developed competing definitions for what they would consider lethal autonomous weapon systems.

The most commonly referred to the definition of lethal autonomous weapon systems originated in a 2012 US Department of Defence directive on autonomous weapon systems. This directive outlined the US Department of Defense's view on developing autonomous capability for weapon systems and the level of human involvement required. This document defines a weapon as fully autonomous if when activated, it 'can select and engage targets without further intervention by a human operator'. Interestingly, Directive 3000.09 lists a requirement for sufficient training for human operators, which indicates a recognition that human operators would have to retain some level of oversight over any use of force decisions. The concern of how to balance the need to achieve effectiveness in a battlespace characterised by an operational tempo that is potentially beyond the capacity of human reaction time, while also maintaining sufficiently the effective human oversight to guard against the unintended engagements[2] is apparent in this directive. Finally, Directive 3000.09 also contained a built-in process for obtaining waivers for development, deployment or even the transfer of, lethal autonomous weapon systems in situations that potentially contravene the policy (Department of Defense, 2012). Despite being due to expire at the end of 2017, Directive 3000.09 was still in effect at the time of writing and features prominently in the developing discourse on LAWS.

As the most commonly cited state definition for autonomous weapon systems, the Directive 3000.09 definition has been used as the starting point for the definitions used by multiple other actors, including non-governmental organisations (NGOs) (such as the Campaign to Stop Killer Robots). While this definition has found traction among scholars, it has largely been received critically. For example, Heather Roff criticised the DoD definition because the terms *select* and *engage* are open to interpretation (Conn, 2016). Rebecca Crootof emphasised the weapon's ability to process information to make targeting decisions,[3] while Michael Horowitz emphasised the ability to select a target that was not pre-selected by an operator. Notwithstanding scholarly critique, the DoD definition arguably remains the natural starting point for developing a working definition of autonomous weapon systems.

Despite its flaws, the US DoD definition does represent a more realistic, if non-specific, view of autonomy in weapon systems than the definitions adopted by some other states. The UK Ministry of Defence definition, for example, refers to autonomous systems having the capability to understand 'higher level intent and direction' and that individual actions 'may not be' predictable. This seems to indicate that a platform or military system must possess AI with a level of self-awareness that bleeds into the field of general AI. It is highly unlikely that any state actor would countenance the development of weapons that they could not predict, even if it were technologically possible to create LAWS with the capacity to interpret higher-level intent. The concept of this level of full autonomy has been justifiably dismissed as a distraction in the literature (Jenks, 2016), as an approach driven by this definition simply does not account for the weapon systems that are actually in development.

On 14 April 2018, China became the first permanent member of the Security Council to publicly endorse a ban on the use of lethal autonomous weapon systems (Kania, 2018). This surprise announcement was initially seized on as a victory by the Campaign to Stop Killer Robots and covered extensively in the media, but closer analysis identifies this announcement as an important example of how states can utilise definitional factors to gain influence over the development of LAWS as an emerging disruptive military innovation.

The Chinese definition of lethal autonomous weapon systems is based around five characteristics, which serve to exclude other forms of increasingly autonomous military technologies from the discourse. The first characteristic is that a device must carry a 'sufficient payload' and be intended to employ lethal force. While this would obviously cover LAWS that are designed to directly participate in combat, it would exclude those that carried a less-than-lethal munitions package (such as the remote-operated 'Skunkcopter' UAV) or are designed for an anti-vehicle/munitions primary function. The second characteristic is an unusually high autonomy barrier, stating that a LAWS would have an 'absence of human intervention and control' for the 'entire process of executing a task'. China's statement was vague about what it considers a 'task', this document could refer to a single use of force decision, the acquisition of a target or an entire deployed mission. Thirdly, and closely linked, the device should have no method of termination once activated to be considered a LAWS. This would discount weapon systems that operate autonomously but can be overridden by a human overseer, such as the Phalanx Close in Weapon System. It is also highly unlikely that a state would deploy a weapon they had no way of deactivating or assuming control over, especially given the comparatively nascent state of artificial intelligence technology.

The fourth characteristic is that the device must have an indiscriminate effect, that the device would 'execute the task of killing and maiming regardless of conditions, scenarios and targets'. This is an interesting inclusion because international humanitarian law already forbids the use of weapon and weapon platforms that are incapable of being operated in a discriminate manner. The inclusion of this characteristic is complemented by the latter statement in the same announcement that a fully autonomous weapon system would be incapable of satisfying the legal requirement of discriminate use of force. The question of whether a fully autonomous platform could abide by international law in the use of discriminate force is central to the debate surrounding LAWS and has been at the forefront of publicly visible developments in the space. As an example, the Super-Aegis II is capable of distinguishing between uniforms and offers clear warnings before engaging to reduce the chances of using lethal force against civilians. Finally, the Chinese definition includes the characteristic that LAWS would be able to evolve and learn through interaction with the environment they are deployed into in such a way that they 'expand its functions and capabilities in a way exceeding human expectations'. This final characteristic leans closer to the UK's definition of fully autonomous weapons and is effectively arguing that the presence of an actively evolving artificial intelligence is necessary for a weapon system to be considered a LAWS. The concept that LAWS are being developed with high-level artificial intelligence has been widely criticised by scholars and defence personnel but is a common point raised by concerned NGOs and smaller states. While it is possible, it is beyond the realm of current technology and whether states would even be interested in a learning autonomous weapon has been criticised as unrealistic.

There are many reasons that the Chinese definition of lethal autonomous weapons is particularly important. Aside from their obvious influence as a permanent member of the Security Council, autonomous military technology is emerging as a key force-multiplier, a factor that is of obvious importance in the context of the Sino-American rivalry and Chinese military modernisation. Furthermore, China has a proven track record of using and then ignoring international law as a tactic for advancing their interests, as an example consider China's reaction to being ruled against by the UN permanent court of arbitration in its case against the Philippines over territorial disputes in 2016 (Zhou, 2016). Finally, China has already emerged as a major exporter of unmanned aerial vehicles (armed and unarmed) to both state and non-state actors (Ewers et al., 2017). Indeed, the 2017 decision to reduce export restrictions on US companies was partially motivated by a desire to counterbalance the market dominance achieved by China in the UAV export market. While China's decision to support a ban on the development and use of autonomous weapon systems seems to be a victory for those opposed to LAWS, the actual content of their announcement reveals the importance of the definitional agreement.

The Chinese announcement clearly excludes large aspects of the developing autonomous military market; however, it has proven quite common in the definitional debate for state and scholarly actors to put forward definitions that have additions that limit the scope of their application. The inclusion of 'Lethal' in

LAWS excludes weapon platforms that are designed to utilise less-than-lethal ammunition or guide other munitions, while the requirement of 'higher level' autonomy excludes the plethora of human supervised weapon systems that are already deployed or in development. As encountered by the UN-sponsored Group of Governmental Experts on LAWS, this disagreement on a common definition hampers efforts to develop either a ban or effective regulatory controls (Human Rights Watch, 2019).

It appears, therefore, that the most effective way to analyse the impact of autonomous weapon systems is to link their definition to a functional assessment of the level of independent control the platform has over its 'critical functions'. The critical functions of a weapon system are the processes used to select, acquire, track and attack targets (ICRC, 2014). These processes are considered critical because they become the core of the kill chain[4] once human supervision is removed (Cheater, 2007). The level of control over these functions is central to the ICRC definition of autonomous weapon systems.[5] This book adopts a functional assessment-based definition of fully autonomous weapon systems. However, a weapon platform that satisfies this definition would not be readily available to all states and non-state actors, nor necessarily would it be the preferred version of this innovation for every actor. Therefore, this book adopts the commonly utilised functional categories to identify three distinct types of autonomous weapon platforms. While the fully developed RMA would be a fully autonomous weapon platform capable of employing lethal force, as with prior innovations, this should not blind the researcher to considering the impact of the adoption of closely related, albeit less advanced, versions.

The complex definitional debate surrounding the term lethal autonomous weapon system is one of the key reasons that international efforts to implement a pre-emptive ban have stalled. Bodies that are believed to be developing autonomous military technology include: the United States, South Korea, China, Russia, India, the United Kingdom, the European Union and Israel, though none has admitted to possessing a functioning fully autonomous weapon system (ICRC, 2016). Only 19 countries publicly support an outright developmental ban, however, this support is based on divergent conceptual understandings of 'fully autonomous weapons'. The clear majority of the 63 other states that have publicly stated a position support the continuation of governmental discussions (Campaign to Stop Killer Robots, 2017). This shows that, while the majority of states do not support a pre-emptive ban, they are concerned and willing to continue high-level discussions towards generating a normative and legal framework to control the impact of LAWS. Outside the land of government press releases, the 2017 inter-governmental meeting of experts was cancelled, ostensibly due to a lack of funds. The 'discussion' advocated by the majority of states this year has therefore been largely organised by non-governmental organisations, scholarly communities and regional inter-state bodies. The development of autonomous military technology has not comparably slowed during this process, bringing us closer to the introduction of fully autonomous military technology without an effective normative framework to govern its impact.

2.3.2 Assumptions of LAWS development

For the purposes of analysis, this book makes four assumptions about the first- and second-generation LAWS. This book focuses on the first- and second-generation LAWS. All major military innovations have evolved over time following their initial impact period. It is neither possible to know how autonomous weapon systems will look in 50 years nor as useful as an examination based on current or in-development technologies and operational concepts. The first assumption is that autonomous military technology development will continue to focus on platforms, rather than munitions or completely independent systems. Fully autonomous weapon systems would necessitate removing human control on a strategic level, something that Southeast Asian states are unlikely to be interested in pursuing due to the strategic risk. Equally, however, because munitions are designed to be expendable, limiting their return on investment, there would be comparatively little advantage in developing autonomously operating munitions rather than supervised munitions. Weapon platforms are a good focus given that LAWS are intended to make tactical decisions to use force and are designed to be re-useable (Horowitz, 2016).

Secondly, it is assumed that no state will deploy a weapon system that completely separates humans from the decision to employ lethal force, at least in a ground-based role. The caveat to this assumption is that, to be effective, LAWS are likely to have control over the immediate release of force, with human operators placed further back in the kill chain. Therefore, this book assumes that land-based platforms will be primarily 'Human on the Loop' systems, while sea- and air-based systems will be closer to the US Do D definition of a fully autonomous weapon. This assumption is supported by the majority of country position statements to the first two international Meetings of Experts on LAWS and the literature.

Thirdly, this book assumes that non-proliferation controls comparable to nuclear weapons will not be imposed, or at least will be ineffectual, in the case of lethal autonomous weapon systems. This is primarily due to the nature of software-based technology, which is inherently more vulnerable to duplication or proliferation. Furthermore, this book assumes that in the absence of such controls, states will be willing to export complete autonomous weapon platforms to friendly states. This could occur in multiple ways, including traditional arms export agreements, co-development agreements or technology exchanges. A key component of this assumption is that private defence firms will be allowed to market weapon platforms of varying levels of autonomy to Southeast Asian states. Based on the current advertising by large defence contractors in the United States, this is clearly an assumption that is shared by the defence industry.

This book further assumes that LAWS will not become of sufficient complexity and security that they cannot be mass-produced by developing states or replicated by smaller states and non-state actors. This assumption is based on the current diffusion and proliferation of remote-operated military systems, especially remote-operated combat aircraft. This assumption is also supported by the recent history of cyber-weaponry being stolen or replicated by non-state actors and smaller states. While the WannaCry attack in May 2017 was initially linked to North Korea, Microsoft subsequently blamed the United States National Security Agency, claiming that the underlying exploit had been stolen from the agency's stockpile of cyber weapons. In short, this book assumes that LAWS will not be of sufficient complexity that the underlying technology does not diffuse in Southeast Asia.

2.4 To ban or not to ban – a question of international law and regulation

At the centre of the burgeoning discourse around autonomous systems is a debate on how lethal autonomous weapon systems would interact with international humanitarian law (IHL), more commonly known as the laws of war. This debate has split scholars and ensured that the vast majority of published works have remained focused on whether IHL can effectively regulate autonomous military technology. At the centre of this debate is the question of whether a pre-emptive ban on the development of autonomous military technology is warranted.

Two major camps have formed in the scholarly community: those in favour of a ban, who are supported by multiple NGOs and those against a developmental ban. The former argue that AWS violate international humanitarian law and international human rights law. Academics who oppose autonomous weapon technology include Asaro and Sharkey. Large NGOs and the former UN Special Rapporteur on Extrajudicial Killings have also published calls for a ban on the basis of ethical, moral and legal objections to '*killer robots*' (Heyns, 2013). At the forefront of the drive for a pre-emptive ban is an NGO, the Campaign to Stop Killer Robots, extremely active advocates who have amassed support from large swathes of the academic and business community (Sample, 2017). This group also keeps a list of country positions on lethal autonomous weapon systems that identifies states who are in favour of a developmental ban.

A smaller, but still substantial, body of scholarly work argues that a pre-emptive ban would not have the impact suggested by advocates. Those scholars who oppose a ban argue that a ban would be ineffective (Schmitt, 2013), that the use of LAWS is sufficiently regulated by existing international laws and norms (Anderson, Reisner and Waxman, 2014), or that, as weapons with some level of autonomous capability already exist, the international community should be focusing its efforts on developing effective regulations as opposed to pursuing a blanket ban (Crootof, 2014). Among opposing scholars, the underlying logic is that responsible design and deployment within existing IHL and other normative frameworks is the most effective way to regulate the impact of lethal autonomous weapon systems. As an example, Kastan (2013) argues that, while a ban is unnecessary, specialised military procedures and adaptions to IHL are needed. This body of scholarly thought is more closely aligned with my perspective.

Advocating for a pre-emptive ban on autonomous military technology requires one to wilfully minimise or ignore the dual-use, software-based nature of its enabling technologies. It also requires that one discount the fact that no weapon system currently exists (based on public knowledge) that crosses the line between 'highly automated' and 'autonomous', although admittedly there remains no universal agreement about where to draw that line or even how to objectively measure the autonomous capability of a given platform.

Furthermore, even if we were to ignore the limitations of current technology, it is difficult to support the related argument (Anderson, 2016) that existing legal weapon review processes would be insufficient for evaluating whether autonomous weapon systems are a legal method (or tool) of warfare, which is distinct from whether a particular LAWS is deployed in a manner consistent with the principles of IHL.

Article 36 of Additional Protocol I of the 1949 Geneva Conventions already requires that states conduct a formal legal review before the procurement of any new weapon system to determine whether it inherently offends IHL, as well as the risks posed in the event of misuse or malfunction (Geneva Academy, 2014). As early as the April 2016 Convention on Certain Conventional Weapons (CCW) Meeting of Governmental Experts on LAWS multiple states publicly agreed that, as with any new weapon system, LAWS should be subject to legal review. It is not unusual for states to alter their process for conducting legal weapon reviews following the emergence of novel or evolutionary weapon systems (Anderson, 2016). Australia presented a detailed description of its System of Control and Applications for AWS (which included legal review) as part of its submissions to the August 2019 meeting of the CCW Group of Governmental Experts on LAWS.

Evaluating whether an emergent weapons system is a just method of warfare, the core purpose of an Article 36 review, relies on three principles, none of which are necessarily offended by shifting the decision to identify and engage targets to a machine. Firstly, the weapon system cannot inherently cause severe environmental damage. Secondly, the weapon must not be indiscriminate,[6] incapable of differentiating between targets. Cluster munitions and biological weapons are examples of indiscriminate weapons. Importantly, this standard applies to the armament itself rather than the identity of the weapon's user (Schmitt and Thurnher, 2012). Therefore, so long as the armament is not indiscriminate (as cluster munitions) then whether the delivery platform is manned, remotely operated or autonomous is not a determinant factor in an Article 36 review.

Thirdly, the humanity standard holds that belligerents do not have an unlimited right to adopt means to injure the enemy and thus weapons cannot inherently cause unnecessary suffering. This standard bans weapons that are 'of a nature' to cause superfluous injury or unnecessary suffering (Pilloud et al., 1987). Blinding lasers and exploding small arms ammunition both violated this standard. Merely controlling a weapon platform remotely (drones) or enabling it to make targeting decisions independently (LAWS) would not influence whether it inherently violates this standard (Martin, 2015). This standard merely requires that belligerents do not inflict unnecessary suffering in pursuit of a military objective. There is no evidence that (for example) a drone strike would cause significantly more direct injury or suffering than the equivalent manned strike or traditional artillery bombardment.

Overall, the argument that existing legal review processes are insufficient in the case of increasingly autonomous weapon systems, or that LAWS inherently violate international humanitarian law does not reflect the focus of these standards, nor that the majority of (publicly acknowledged) unmanned systems (remote-operated, highly automated or even with limited autonomy) are generally platforms that carry legacy weaponry that has undergone previous legal review. For example, the South Korean Super-Aegis II (referred to in the Introduction) is equipped with a 12.7-mm machine gun, versions of which have been regularly deployed by various militaries over the past 60 years.

Whether delegating the decision to end a human life to a machine would be ethically justifiable or not, while an important question, is not considered by these standards. Instead, some advocates of a pre-emptive ban have argued that these ethical concerns would be sufficient to violate the Martens Clause,[7] drawing parallels to the ban on blinding lasers, arguing that they also violated the principle of public conscience. Despite being an ongoing point of contention in the literature, it is difficult to evaluate the applicability of the Martens Clause simply because there is a dearth of large-scale studies of public opinion towards the increasingly autonomous weapon systems.

Based on the available evidence, it seems clear that armed drones and LAWS are a legal method of warfare. However, ongoing legal reviews of individual emerging weapon systems are essential to ensure that new models do not individually violate these standards. Even when inherently legal as a method of warfare, weapons must be utilised in a manner that is consistent with the IHL principles of proportionality, necessity, distinction and precautions in attack.

The principle of proportionality establishes that belligerents cannot launch attacks that could be expected to cause a level of civilian death or injury or damage to civilian property that is excessive compared to the specific military objective of that attack. Attacks that recklessly cause excessive damage, or those launched with knowledge that the toll in civilian lives would be clearly excessive, constitute a war crime (Dinstein, 2016). The test under customary international law applies a subjective 'reasonable commander standard' based on the information available at the time. To be deployed in a manner that complies with IHL, an autonomous platform would require the ability to reliably assess proportionality. Current generation AI is unable to satisfy a standard that was designed and interpreted as subjective, although this could change as sensor technology develops (Arkin, 2008).

The principle of military necessity reflects the philosophical conflict between applying lawful limitations to conflict and accepting the reality of warfare. The principle of military necessity requires belligerents to limit armed attacks to 'military objectives' that offer a 'definite military advantage' (Martin, 2015). Furthermore, attacks against civilian objects and destruction or seizure of property not 'imperatively demanded by the necessities of war' are considered war crimes. This principle cannot be applied to a particular weapon platform as a whole; rather it must be considered on a case-by-case basis (Martin, 2015).

The principle of distinction requires belligerents to distinguish between combatants and non-combatants as well as between military and civilian objects (including property), and is the most challenging principle for a military to utilise LAWS in accordance with. At its core, an autonomous weapon system is a series of sensors feeding into a processor; interpreting data to make an active identification and evaluation of a potential target (Stevenson et al., 2015). This is distinct from an automatic weapon, which fires once it encounters a particular stimulus, such as an individual's weight in the case of landmines. The technology does not currently exist that would allow LAWS to reliably identify illegitimate targets in a dynamic ground combat environment. A deployed LAWS would need a number of features including the ability to receive constant updates on the battlefield circumstances (Geneva Academy, 2014); recognition software to recognise the difference between combatants and non-combatants as well as between allies and enemies in an environment where neither side always wears uniforms; and the ability to recognise when an enemy combatant has become *hors de combat*. There are too many variables on the modern battlefield, particularly in a counter-insurgency operation, for any sort of certainty that autonomous weapons will always make the same decision.

Overall, it is insufficient to push for the imposition of a development or deployment ban under IHL on an innovation that has not yet fully emerged. Beyond its questionable practicality, this push has become so central to the discourse surrounding LAWS that it is stifling progress towards arguably more effective outcomes such as: a standard function-based definition; a stronger understanding of the technological limitations among policymakers and end-users; changes to operational procedures to improve accountability; or standardising the benchmarks for Article 36 reviews of AI-enabled weapon platforms.

2.5 How to ensure accountability in autonomous weapon systems

The second theme in the scholarly literature examining autonomous military technology is the practicality of programming normative frameworks (specifically IHL) and ethical behaviour controls into otherwise fully autonomous weapon platforms. The main proponent of installing ethical controls into autonomous weapon systems was Arkin, who first proposed the use of 'ethical governors' in 2008. Arkin's governors would consist of decision gateways, based on basic IHL principles, which a LAWS would progress through in determining whether to use force (Arkin, 2008). Comparing these governors to written rules of engagement for human soldiers would be a simplified but effective conception. Alternative methods proposed in the literature include facial and pattern recognition technology, predictive behaviour systems and advanced machine learning (Sharkey, 2010). In opposition to these proponents there are academics such as Wagner and Sparrow. The former argues that LAWS are fundamentally incapable of abiding by the spirit of IHL and ethical conflict. This is partly due to the subjective nature of applying the proportionality principle but he also points to the ethical problem of LAWS being physically incapable of compassion and empathy (Wagner, 2014), an issue that has also been raised by Sparrow (2015). The core contention of these works is whether there is a technological solution to the practical and ethical problems with giving robots the ability to autonomous use lethal force. Although this book does not directly contribute to this debate, it does integrate the work of roboticists such as Arkin, the concerns of ethicists like Wagner and a technical engineering perspective into its theoretical approach.

While the concept of putting a robot on trial is farcical, there has never been a need for the international community to contemplate accountability when the violation cannot be directly attributed to a human. This question of liability is another key theme in the literature, although it is mostly occurring in scholarly works and publications by think tanks, with organisations and states staying relatively quiet. The question of how to determine liability for the actions of autonomous weapons stems from one of the earliest supervised autonomous weapon system (the Aegis Combat System),[8] which was involved in the downing of the Iran Air Flight 655 (Kastan, 2013). The challenge behind this question is simply that LAWS are robots, non-sentient objects that cannot be held accountable for violations of international law; to further the above example, imagine the ludicrousness of jailing an autonomous weapon. The leading academic proposals in response to this issue focus on combining extended forms of command responsibility and commercial product liability. Margulies' dynamic diligence approach is an extension of the former (2016), while Crootof focused on the latter (2016). In response to this gap, the international community has seized on the concept of Meaningful Human Control, a concept that has no commonly accepted meaning but seems to be accepted by the majority of actors.

The concept of *Meaningful Human Control* arose as a response to this 'accountability gap' (Krishnan, 2009) and has been a major talking point at each meeting of experts. Despite its prominence in the literature and government policy, there is no universal agreement on the limits of its meaning. For example, Heyns has previously written that autonomous law enforcement weapons would still be under Meaningful Human Control if a human authorised that specific target and instance of force, even if the weapons did not engage immediately. The literature has begun to push back against this lack of definitional clarity, as well as the murkiness surrounding definitions of autonomy in the military context (Anderson, 2016). As a prominent example, Crootof has challenged the blind acceptance of Meaningful Human Control. Instead, her work explores how the concept of Meaningful Human Control would interact with inconsistent domestic state laws as well as international humanitarian law (Crootof, 2016). In 2016, Horowitz published a meta-analysis of the various definitions of autonomy in the lethal autonomous weapon system space, which included reference to this lack of clarity around Meaningful Human Control.

This book avoids extensive engagement with the issue of accountability for the actions of autonomous weapon systems. This is due to time, space and complexity limitations, which are particularly prevalent given its focus on Southeast

Asian regional security. This book does, however, engage with and evaluate the impact that such definitional disagreements have on the counter-proliferation efforts when dealing with the emergence of Revolutions in Military Affairs. This book also explores the value of common definitions and legal inter-compatibility between regional actors as they establish a new normative framework in the wake of the emergence of a disruptive military innovation, both historically and currently with autonomous weapon systems.

While the majority of the existing scholarly work relates to either the legal, ethical or moral consequences of developing and deploying autonomous weapon systems, there are also scholars who argue that the development of lethal autonomous weapons will reduce the impact of war on civilians, protect the lives of soldiers and minimise the brutality of war. The underlying current of their argument is that the widespread diffusion and proliferation of autonomous military technology could bring about 'sterile' or 'bloodless' warfare and that there is therefore a 'moral duty' to develop autonomous military platforms. Proponents of this view include Strawser (2010) and Lucas (2014). It is noteworthy that similar views were expressed about other disruptive military innovations.[9] My view of this argument is that it is fundamentally flawed. Given the historical precedent, it appears far more likely that the development of LAWS will merely provide states the ability to persecute armed action without risk to the aggressor. While lives will be lost, they will only be from the targeted community (Steuter and Wills, 2009). This book critically engages with that belief and examines the security impact of smaller states gaining access to autonomous weapon systems. If proponents like Strawser were correct, then the following analysis should support their contention.

A closely related debate that is occurring principally among scholars whose main interest is ethical in nature centres on the asymmetry objection. This objection basically maintains that it is inherently unjust for a state to use vastly superior technology to inflict damage without any risk to its own personnel (Galliott, 2012). While this is of obvious relevance to LAWS, similar objections have been raised in the context of other military inventions that increased the moral and physical distance between a combatant and their target. Kahn goes a step further, arguing that a conflict must involve a level of mutual risk to be objectively 'just' (Kahn, 2002). This is supported by Chamayou, who has referred to drone warfare as 'cowardly and contemptuous' (2015) and argues that by removing the combatant from the risk of physical harm, warfare is no longer a contest; rather it is closer to the application of state force in criminal prosecution (2011). Schmitt (2013), Whittle (2014) and Lucas (2014) challenge this position, arguing that there is no obligation for states to restrain their technological advantage to ensure 'fairness' in the actual conduct of warfare. This view is consistent with public statements by senior US military figures along the lines of 'We're not interested in a fair fight with anyone'.[10] While sections of this book make a general contribution to this debate, due to space and time constraints it is not a major focus.

2.6 Considering public opinion towards autonomous weapon systems

A relatively poorly covered theme in the literature is a lack of non-US data to determine public opinion towards autonomous weapon systems. To date, there have only been three studies (two were US-based). Chronologically, Carpenter conducted the first study in 2013. It showed that 55% of respondents opposed autonomous weapons (39% strongly opposed). Although its methodology was flawed,[11] it is still widely referenced in official documents and the academic literature.

The second US study drew on two experiments conducted by Horowitz in 2015. Horowitz found that the baseline level of opposition to autonomous weapons dropped from 48% to 27% if autonomous weapons protected the US soldiers and were more effective than remote-operated weapons (Horowitz, 2016). While Horowitz's study was focused on the United States with a relatively limited respondent base, it does indicate that there are circumstances in which public opposition is diminished. These studies provide an initial level of insight into the US public opinion of autonomous weapons and have already been used to inform the international debate.

The Open Robo-ethics Initiative (ORI) conducted the only non-US study in November 2015.[12] The results of this survey were fairly clear with 85% of respondents saying that LAWS should not be used offensively and 67% supporting a ban. The most common reason for opposing LAWS was that only humans should be allowed to make the decision to end life. Unfortunately, this survey did not meaningfully engage with the non-English speaking world, one of its stated aims. Although 11.6% of respondents were from South Korea, only two other non-western states had more than ten respondents, Mexico (7%) and India (1.9%) (Open Robo-ethics Initiative, 2015). This leaves a crucial gap in public understanding, particularly because South Korea and India are leading developers of LAWS.

More recently, there were two quite limited surveys commissioned by the Campaign to Stop Killer Robots, the first in 2017 and the second in 2019. While these surveys included participants outside of the United States, no ASEAN member states were among the surveyed countries. These surveys found that opposition to autonomous weapons rising, hitting 61% in the second survey. However, in addition to not including Southeast Asian states, these surveys were quite limited in scope, with only those who indicated opposition being asked the survey's second question. Finally, the data from these surveys is of questionable value for actually informing policy beyond supporting a general call for a ban, given Horowitz's (2016) findings that the composition of the question was influential when measuring public reaction to LAWS.

This gap in understanding public perception of autonomous military technology extends to Southeast Asia. There are currently no scholarly publications available that examine Indonesian or Singaporean public opinion towards autonomous weapon systems. This is a major gap in the literature that could undermine domestic impetus for timely development of regulation. Addressing this literature gap in this book is not feasible due to the required scale, time commitment and resource cost. However, it is a key area for further research, particularly in the Southeast Asian context.

2.7 Lethal autonomous weapon systems, artificial intelligence and international stability

Autonomous military technology has the capacity to fundamentally change how states conduct hostilities and, once it matures, will be relatively inexpensive compared to equally advanced military technology. However, despite the growing body of literature examining the proliferation of remote-operated systems (specifically UCAVs), there is little literature available that examines the impact that the diffusion of autonomous military technology will have on international security and the balance of power.

Published scholarship focusing on the security impact of AWS and AI proliferation or diffusion includes several pieces by Horowitz, a research paper published by the Centre for a New American Security, a special issue of the *Journal of Strategic Studies* published in August 2019, and an earlier article written by Altmann and Sauer (2017). While the recent uptick in scholarly interest is encouraging, these research efforts do not explicitly focus upon the expanded potential role of middle power states in the incubation and early post-demonstration point period of this innovation, nor do they engage with Southeast Asia.

As an illustrative example, consider the opening set-piece in Horowitz's (2019) contribution to the recent *Journal of Strategic Studies* special issue. Utilising the example of the Cuban Missile Crisis, Horowitz illustrated how the presence of autonomously operating weapon systems would have potentially increased the risk of escalation and undermined the United States' effort to deter the Soviets from attempting to run their blockade. This was a valuable and well-reasoned argument, presented through a well-known example of an international crisis. An alternative use of this example, however, would be to consider the impact if Cuba either interfered with the United States–Soviet Union standoff with unmanned platforms, or independently triggered an escalation with difficult to attribute AWS. As Barkawi and Laffey point out in their 2006 article, far from being subordinated to the will of their superpower ally, the Cuban government was a key influencer of Soviet behaviour during the Cuban Missile Crisis. The levelling effect of increasingly autonomous weapon systems gives smaller powers a greater level of agency in great power conflict and competition, a factor that has not been adequately considered in the literature. This gap is the main focus of this book, asking what happens if the modern Melians can capitalise on artificial intelligence and autonomous systems to offset the traditional power dominance of their great power patrons.

The rise of autonomous military technology has sparked major scholarly and governmental interest in the prospect of lethal autonomous weapon systems. This interest has translated into a young but rapidly growing body of scholarly, governmental and technical literature. This body of understanding suffers from being segmented along discipline lines, which this book crosses with an interdisciplinary approach. The key areas of scholarly understanding this book contributes to, are the security impact of LAWS diffusing to minor states and non-state actors, the debate around whether to implement a ban on development and the emerging understanding of the importance of a common definitional foundation for further regulation of this emerging technology. This book's main contribution, however, is integrating the scholarly understanding of autonomous military technology with military diffusion and innovation theory to more accurately estimate the initial impact of the demonstration of lethal autonomous weapon systems as a complete and disruptive Revolution in Military Affairs.

2.8 Military innovation and diffusion theory

Historically, the development of new, paradigm-shifting technologies has been one of the main methods of increasing human influence over the natural environment. Nowhere has this been more apparent in the development of new ways to inflict violence and exert power. Within the commercial world, technological or process innovations that force structural change in the market by upsetting the orthodox market wisdom are known as disruptive innovations (Christensen, 2015). The development of smartphones is an oft-cited example of a disruptive technology that has had widespread economic, social, political and military impacts.

The closest military equivalent is disruptive military innovations, although other terms have been used. These are military innovations that disrupt the existing paradigm of human conflict and upset the pre-existing balance of power. Certain innovations have historically been significant to contributing factors in enabling or triggering hegemonic conflicts that marked the rise and fall of great powers. This section will summarise the major themes in the large body of scholarly work that exists which engages with military innovation and the diffusion of new technology. It will identify and evaluate key theories and extract relevant pieces to establish the theoretical framework that underpins this book's approach to understanding the impact of innovations that undermine transitions of international power.

Despite the popular assumption of militaries as slow-moving, stagnant organisations that are adverse to change and change-makers (which have admittedly generally proven accurate), there is also a long list of major innovations that started their development with military funding; arguably the most famous being computers and the internet. While military bureaucracy certainly can stifle innovation (Grissom 2006), the reality is that advanced militaries (particularly the United States, the Republic of Korea and Singapore) rely upon their technological superiority over rivals to deter aggression and project stability.

There is a large body of literature available that explores how military innovation occurs. Grissom (2006) presents a useful summary of the key theorists and theories. Within this body of literature, there are a number of divergent understandings of how to categorise and understand military innovation. Scholars who have proffered particularly useful definitions of major innovations include Rosen,[13] Grissom[14] and Horowitz.[15] Common across these definitions is an acknowledgement that invention must be combined with change to the 'operational praxis' (Grissom, 2006) to become an innovation, a process that includes the creation of a new strategic doctrine that enables the state to capitalise on the technological invention. The same would also be true in reverse. Without the complementary component, the innovation cannot be considered 'complete'. These two components are often referred to as 'hardware' and 'software', respectively. The former refers to the physical invention or advancement, while the latter refers to doctrinal, operational and organisational change.

Once both factors have matured, it is only a matter of time before the complete innovation is deployed or acknowledged publicly. This is referred to as the demonstration point, after which rival states are faced with the choice of whether to adopt the innovation in question or accept a resulting shift in the balance of power. However, there is no guarantee that the demonstration point will occur shortly after both components are developed, and in some cases, the technology has already begun to mature before a novel operational concept emerges or a war begins, which in turn triggers a demonstration point. As an example, consider armoured warfare, as an innovation it was only completed by the emergence of German armoured warfare doctrine; combining aircraft, logistics, radios and combined arms manoeuvre with the armoured vehicles themselves (operational praxis) in the early 1930s and demonstrated in 1939. This was over 20 years after the first deployment of the modern tank (the invention component) by the British (Horowitz, 2010), who had instead developed tanks that were designed for the previous war's battlespace (infantry tanks such as the Matilda I) or reflected a distinctly naval view of utilising columns of comparatively fast and lightly armoured tanks for independent penetration operations (Cruiser Mk1); both of which proved inferior 'software' components.

Applying disruptive innovation theory to military technological development is not unprecedented. The *Revolution in Military Affairs* theoretical framework initially appeared during the Cold War and was popularised during the 1990 Gulf War (Galdi, 1995). RMA refers to a drastic alteration of the nature of armed conflict due to the development, or innovative application, of a disruptive new military technology. Importantly, an RMA is a complete innovation, combining a disruptive invention with drastically altered military doctrine or organisational change (a typology of military innovation varieties is outlined below in Figure 2.1).[16] The result is an innovation that disrupts the enduring character of warfare, fundamentally altering the character and conduct of military operations. Similar to Horowitz's Major Military Innovations (2010), the literature on RMAs has typically given priority to leading military powers, while smaller powers, lacking the resources to become competitive early adopters, instead undertake to take alternative responses, such as bandwagoning or re-asserting neutrality.

The core of the definition of RMA is that they are radical innovations that constitute discontinuities in military affairs (Vickers, 2004). The inevitable

HARDWARE
(Weapon/platform/system)

	Incremental	Discontinuous
SOFTWARE (Doctrine/organization) — Incremental	SUSTAINING INNOVATION	TECHNOLOGICAL BREAKTHROUGH (Weapon/platform/system)
SOFTWARE (Doctrine/organization) — Discontinuous	ARCHITECTURAL BREAKTHROUGH (Doctrine/organization)	DISRUPTIVE, REVOLUTIONARY INNOVATION

Figure 2.1 Matrix of innovation types

disagreements in the scholarly community over which innovations can be considered as RMAs and which are simply major innovations of the kind envisaged by Grissom and Rosen are still ongoing (Vickers, 2010). Key theoretical approaches for determining whether an innovation is an RMA include Krepinevich's Technology-Concept-Organisation Theory (1992) and Boot's significant Four Revolutions argument (2006). This debate has maintained a historical focus and is largely avoided as not directly relevant to this book. However, Table 2.1 presents the extensive list of innovations that have been referred to as RMAs, as compiled by Vickers (2010). As with Vickers, this book presents this list merely as an illustrative framing object, and does not assert that this is an exhaustive or universally agreed list of paradigm-shifting military innovations. It is important to note that RMA as a theory remains heavily contested, particularly in more recently published literature. Criticism typically centres on its techno-centrism or the assertion that it fails to adequately account for enemy action (Futter and Collins, 2015). However, this is not a one-sided debate and scholars such as Raska (2020) continue to analyse the emergence of AI-enabled systems through the lens of RMA. While it is important to acknowledge and account for these criticisms, this book adapts elements of Revolution in Military Affairs into its overarching theoretical framework.

A comparative weakness remains in the literature in applying military innovation and diffusion theories to lethal autonomous weapon systems. Contributing to this limited scholarly understanding of autonomous military technology as paradigm-shifting innovation is a key contribution of this book.

In addition to the available literature and scholarly work that explores the nature and definition of various types of military innovation, there is a related body of scholarly work that focuses on exploring how and why such innovations diffuse, with a focus on the factors that influence individual states to acquire major military innovations. The fact that the key enabling technology for LAWS (Artificial Intelligence) can be comparatively easily replicated should be a major concern

Table 2.1 List of Innovations Referred to as RMA

Revolution in Weapons Technology	Military Revolution of the Ch'in	Gunpowder Infantry (Spanish)	Railroad, Rifle and Telegraph	Radar
Advent of Bronze Weapons	Greek Fire	Fortress Revolution	Ironclad Warships	Amphibious Warfare
Professional Warriors	Shock Cavalry (Stirrup)	Dutch-Swedish Tactical Reforms	Battleship-Battle Cruiser	Signals Intelligence
Emergence of Chariot Warfare	Mongol Swarming Tactics	Creation of Modern Military Institutions	Submarine	Atomic Weapons
Eruption of Massed Infantry	Longbow	French Military Reforms	Air Warfare	Thermonuclear Weapons
Cavalry Revolution	Offensive-Defensive Strategy	Fleet Battle Line	Second World War Combined Arms	Ballistic Missiles
Specialised Naval Vessels	Swiss Pikemen	Revolution in Military Finance	Armoured/Air Warfare	Nuclear-Powered Submarines
Emergence of Citizen-Soldier	Artillery Revolution	Flintlock/Socket Bayonet and Line of Battle	Carrier War	People's War
Revolution in Greek Battle Tactics	Guns and Sails	French Revolution/Napoleonic	Strategic Bombing	Photo-Reconnaissance Satellites
Macedonian Integrated New Model Army				

to security officials and is a focal point of this book. This book also engages with the role states are likely to play in the proliferation of autonomous weapon systems. The difficulty and major costs of autonomous weapons are the results of increasing reliability, safety and advanced targeting. For violent non-state armed groups, basic AWS will be extremely attractive. This book draws on two theories of innovation and diffusion, which complement each other and inform its theoretical framework.

This book draws primarily on ACT, which argues that, when a major military innovation reaches its demonstration point (generally, but not exclusively, when the complete innovation is demonstrated in a conflict), arguing that the resulting demonstration effect prompts the decision whether to innovate. This decision is then determined by two factors. Firstly, the financial intensity required to adopt a given innovation. This goes beyond a simple examination of the construction/ acquisition cost of an innovative platform in US dollars, although the per-unit cost of a platform remains influential. Instead, it uses a wider definition, referring to 'the particular resource mobilization requirements involved in attempting to adopt a major military innovation' (Horowitz, 2010). For example, it includes an assumption that innovations with dual-use have lower financial intensity due to the contribution of the civilian sector as well as the influence of domestic politics and norms. The second factor is the state's organisation capital capacity, which refers to the flexibility and capacity of a state military to 'respond to changes in the character of warfare' (Horowitz, 2010). Important indicators of organisational capital capacity include the age and complexity of the state, the specificity and relevance of its primary task and its willingness to experiment (Horowitz, 2010). Furthermore, the level of financial intensity and organisational capacity required by a given innovation can have a systematic effect on the rate of its proliferation across related states. The uniquely disruptive nature of LAWS becomes apparent under this theory given the lack of reliance on resource-intense hardware components at the entry level. In the current globalised, information-rich world this means that there would be lower transferability barriers to non-great power states adopting autonomous military technology. This theory is utilised because it draws on a representative range of key indicators, which can be applied to the case-study states.

However, given concerns raised in the literature, that it does not sufficiently consider the effect of international norms or domestic political and cultural influences (Gilli and Gilli, 2014; Saitou, 2012), this book supplements Adoption Capacity Theory with the Organisation theory of military innovation and diffusion, as well as elements drawn from the Revolution in Military Affairs theory. The organisation theory holds that the 'origin, diffusion and influence of a particular invention cannot be understood except in terms of the total culture which originated or utilizes it' (Goldman and Andres, 1999). An organisation theory approach to innovation diffusion identifies three key indicators of the capacity for a state to capitalise on a diffusing military innovation. These are the technological capacity of the identified state, the development of effective military training and doctrine that exploits the advance and the receptiveness of the domestic socio-political and cultural environment to the underlying technology (Goldman and Andres, 1999). Beyond a hybrid of the Adoption Capacity Theory and Organisational theory, this book departs further from orthodox diffusion theory in order to account for the influence of precursor technologies on the diffusion of autonomous military technology.

Precursor innovations impact the development and understanding of truly revolutionary advances. The inclusion of analysis of precursor military innovations is a departure from previous analyses of innovation and military diffusion theory, which generally minimises or ignores the role of precursor technology as a doctrinal bridge for policy and military leaders. Krepinevich (1994) has argued that the Gulf War was a 'precursor war', which fulfilled a similar function. In the case of LAWS, precursor advancements include remote-operated unmanned platforms (principally UCAVs), 'smart' munitions and the doctrinal concept of Networked Warfare (which was the impetus for the initial development of RMA in the literature).

Other established theories of why states adopt military innovations include the Offense-Defence theory advocated initially by Resende-Santos (1996), which argues that when the benefit of an innovation becomes apparent to other states, there is a demonstration effect, which plays into either an offensive or defensive power imbalance, prompting the adoption of the new weapon. A major criticism of this approach is the difficulty establishing definitively whether an innovation is offensive or defensive, this is particularly problematic given the inherently dual-use nature of autonomous technology. Therefore, it is not used in this book.

Beyond theoretical approaches, this book responds to the limited published scholarly literature (in English) that examines the military innovation, diffusion and proliferation process from the perspectives of middle power states and, more specifically, ASEAN member states. Current scholarship examining the military innovation processes of the Singaporean Armed Forces and *Tentara Nasional Indonesia* is limited. Scholars writing in this space include Raska (2015, 2019); Bitzinger (2018); Laksmana (2018); Andrew H. Tan (2013); See Sang Tan (2015); Schreer (2015); and Syailendra (2017). However, none of these authors has engaged directly with autonomous weapon systems, nor is there literature that applies Adoption Capacity Theory to case studies involving Indonesia and Singapore or examines their adoption capacity. This book contributes to the literature by applying military innovation and diffusion theories to Southeast Asian states in the case of LAWS.

As the main component of this theoretical framework, an understanding of how military innovation and diffusion occurs is vital for comparative analysis of the capacity of the case-study states to integrate LAWS. This book harnesses and develops the existing literature to identify historical precedents from prior RMAs that are relevant to an analysis of LAWS. Drawing primarily on Adoption Capacity Theory and organisational innovation theory, the theoretical framework that underpins this book takes a novel approach to understand how RMAs diffuse and proliferate, as well as the impact of this process on regional security environment. This is particularly important given the link between the emergence of RMAs and transitions in hegemonic power (Metz and Kievit, 1994).

2.9 Power transition and hegemonic conflict

The final aspect of the theoretical framework for this book is the role of disruptive military innovations in the transition of hegemonic power. An enduring historical tenet, this refers to the transition of prominence between existing and emerging powers, a transition that is often violent. These transitions are triggered when a rising hegemonic challenger feels suppressed by the existing balance of power and acquires the means to challenge the dominance of the hegemon. The emergence of a disruptive military innovation, such as lethal autonomous weapon systems, has historically been a common enabler of such challenges. Three inter-related theories are applicable to this book's engagement with this process, Power Transition Theory, Hegemonic War Theory and the Thucydides Trap, all of which are viewed through an offensive neo-realist perspective.[17]

Power Transition Theory (PTT) is a cyclical, hegemonic realist approach to international relations, which posits that, although a dominant power is highly influential on the international stage, the underlying balance of power is fluid (DiCicco and Levy, 1999). Therefore, it is subject to change based on the internal growth and development of lower-tier challenger states (Lebow and Valentino, 2009). These challenger states can become dissatisfied with the current balance of power and attempt to instigate change. The larger hegemon eventually loses influence to an energetic, powerful rival and influence over the balance of power shifts to the new hegemon. Hegemonic War Theory (HWT) is an extension of this theory that argues that the transition between the dominant power and the challenger can lead to conflict, perhaps warfare. In the case of military innovation, if a challenger state, which is dissatisfied with the status quo, was able to increase its power by rapidly adopting an emerging disruptive military innovation, it would prompt the dominant state to adopt or improve upon that innovation to re-secure its position. This process generates regional instability and has the potential to trigger a hegemonic war, as the dominant power reacts violently to the transition of power towards the rising power.

Previous disruptive military innovations have precipitated shifts in the ability of states to project their power (Metz and Kievit, 1994). Rising states will capitalise on emerging innovations to secure a power advantage, while smaller states will imitate and emulate the more successful states to secure their own power base from their rivals, increasing the rate of diffusion (Goldman and Andres, 1999). Given the ease of emulating inherently dual-use technologies like autonomous weapons, this diffusion could have a greater role in creating instability than the power transition itself, given the increasingly information-based international order (Nye, 2010). These theories postulate that the diffusion of an RMA precipitates but is not necessarily sufficient to trigger hegemonic war. The resulting power transition will favour the state or non-state actor who most effectively capitalised on the disruptive innovation with a suitable doctrine (Gilpin, 1988).

A modern combination of aspects of this process is evident in the work of the Thucydides Trap project, led by Graham Allison, which examines a series of hegemonic power transitions to determine what factors influence whether a hegemonic conflict erupts into warfare. Allison's book, with its modern focus on the emerging hegemonic tension between China and the United States, is a core text relied upon by this book's theoretical framework (Allison, 2017). Taking this approach as a basis, this book's theoretical framework then departs substantively with the inclusion of minor states and non-state actors into the consideration of potential hegemonic conflict.

Where previous hegemonic conflicts were enabled by RMAs, the high technical and financial barriers to their adoption and limited diffusion of information meant that these conflicts were dominated by large powers with the capacity to adopt the underlying innovation in the immediate aftermath of its demonstration, or, alternatively to rapidly constitute alliances to offset its influence. The end result was classically demonstrated in the Melian dialogue, weaker states were largely subsumed by the requirements of the hegemonic powers. LAWS are unlike other expensive advancements in military technology[18] in their vulnerability to rapid diffusion and proliferation. This is because their underlying technology can be inexpensively copied or stolen. In this manner, LAWS are more comparable to cyber-weapons. This means that autonomous military technology is the first major military innovation that is sufficiently vulnerable to diffusion and proliferation that there is no guarantee that either hegemonic state would be able to maintain sufficient technological superiority to conscript smaller states into their conflict to impose its influence after the conclusion of a multi-sided hegemonic conflict. This uniquely disruptive nature has not been considered by existing scholarly work and is a key contribution of this book.

2.10 Conclusion

The interdisciplinary nature of this book is reflective of the multiple relevant themes of scholarly work that address aspects of the impact of lethal autonomous weapon systems. The theoretical framework provided in this section forms the grounding for later analysis.

The emergence of autonomous weapon technology has sparked a series of cross-disciplinary debates among scholars, engineers and policy analysts. These debates have characterised the available body of scholarly work. The first major debate centres on whether a pre-emptive ban on the development of lethal autonomous weapon systems is necessary or would even be effective. From reviewing papers from either side of this debate as well as NGO and governmental position papers it is clear that, while a ban is not an effective option, there does need to be effective regulation and amendment to the international rules-based order to minimise the negative impact of autonomous military technology diffusion. This book does not deeply engage with the issue of Meaningful Human Control and its uncertain meaning or the ethicality of deploying autonomous weapon systems, while these are noble avenues of scholarly research, time and resource limits prevent their exploration in this book. The main contribution of this book to the scholarly understanding of lethal autonomous weapon systems is integrating this body of work with military diffusion and innovation theory to more accurately estimate the impact of autonomous weapon diffusion to smaller states within Southeast Asia.

At the core of the impetus to develop LAWS is the persistent belief that they represent the dawn of an age of 'sterile' or bloodless warfare. A common characteristic among technology-based military innovations is increasing the physical and psychological distance between the weapon's operator and the target while increasing individual lethality. In theory, LAWS would remove human soldiers from the risk of direct harm during the imposition of state violence. The use of UCAVs over the past decade has reinforced the notion that LAWS, which can be seen as UCAVs' natural successors, will be used as a tool of international power projection and state violence.

The effectiveness of a ban is less important than designing effective policy responses before diffusion occurs on a large scale. As an immediate example, consider the proliferation of remote-operated weapon platforms, which have spread to more than 80 state and non-state groups (Sayler, 2015) including organised criminal groups, terrorist organisations and law enforcement. By integrating elements of distinct theories of military diffusion in its theoretical framework, this book is departing from the theoretical orthodoxy and contributing to the development of a new model for understanding how disruptive RMAs impact the transition of hegemonic power.

This book draws on a hybridised theory of military innovation and diffusion to enable exploration of LAWS as a disruptive military innovation. The theoretical approach utilised draws primarily upon Adoption Capacity Theory and organisational innovation theory, with the inclusion of precursor innovations and elements drawn from RMA theory.

This book will be the first major piece of scholarly work that specifically examines how the diffusion of LAWS will impact relations of power among Southeast Asian states. Within this, it will contribute to the literature across three key existing weaknesses. The first contribution will be to the limited body of literature that examines how Southeast Asian states have responded to the development and proliferation of remote-operated unmanned combat vehicles. Secondly, this book applies a supplemented version of Adoption Capacity Theory to LAWS as an emerging innovation as well as to Southeast Asia as a novel geographic focus, bringing additional attention to the military innovation process of middle power militaries in this region. Finally, this book engages with the key distinction between disruptive military innovations and RMAs, which is how they impact hegemonic transitions of power. Building on the works of Gilpin and Allison, this book examines the role of hegemonic conflict by introducing the concept of disruptive innovations that enable minor states and non-state actors to play a more independent role in the hegemonic conflict. A core contribution of this book is exploring how the involvement of new players in the Thucydides Trap will affect the transition of hegemonic power between China and the United States as well as the security of Southeast Asia.

This theoretical framework will inform and guide the latter book chapters as well as situating their contribution within the wider scholarly literature of each of the three theoretical pillars explored above. The application of this framework will be guided by a mixed methodological approach that emphasises case studies to demonstrate the impact of autonomous weapon systems in this region.

Notes

1 For a comprehensive account of the development of the MQ-1 Predator, see Whittle, 2014.
2 Directive 3000.09 defines 'unintended engagements' as 'The use of force resulting in damage to persons or objects that human operators did not intend to be the targets of U.S. military operations'.
3 'A weapon system that, based on conclusions derived from gathered information and preprogramed constraints, is capable of independently selecting and engaging targets' – Rebecca Crootof quoted in Horowitz, 2016.
4 The *kill chain* is a commonly used term within the US military and in the relevant academic literature. It refers to the targeting process used in air strikes, which comprises of Find, Fix, Track, Target, Engage and Assess (F2T2EA). It is enshrined in US Air Force doctrine and also referred to as the 'Dynamic Targeting' process (United States Air Force Air Education and Training Command, 2017).
5 'Any weapon system with autonomy in its critical functions. That is, a weapon system that can select (i.e. search for or detect, identify, track, select) and attack (i.e. use force against, neutralize, damage or destroy) targets without human intervention' (ICRC 2015).
6 This is different from the principle of distinction, which relates to the way a weapon system is used.
7 The Martens Clause requires that the legality of new weapon systems be subject to the principles of humanity and the dictates of public conscience in cases that are not covered by established international law (ICRC, 2014).
8 The Australian Navy chose to equip its Hobart Class Air Warfare Destroyers with an upgraded version of this platform (Mugg, Hawkins and Coyne, 2016).
9 For example, the inventor of the first mass-produced machine gun (Richard Gatling) believed that his invention would 'supersede the necessity of large armies, and consequently exposure to battle and disease would be greatly reduced' (Chivers, 2010).
10 General Frederick Hodges quoted in Huggler (2015).
11 The first methodological flaw was that the respondents were sourced from an online, rewards based private recruitment firm (YouGov). This method of respondent recruitment has well-known reliability problems, with recorded instances of individuals registering multiple accounts to gain more rewards. Secondly, Carpenter's study is undermined by using leading and highly emotive questions. This is a topic that the general public would know very little about beyond their immediate association of 'robotic weapons' with the *Terminator* movie franchise. Horowitz's findings demonstrated that contextualised questioning is particularly potent in this field.
12 The ORI has left the survey open to continue collecting data, interested readers can view their most up to date results at this address: http://www.openroboethics.org/2015-laws-result/
13 Rosen defines a Major Innovation as 'a change that forces one of the primary combat arms of service to change its concepts of operation and its relation to other combat arms, and to abandon or downgrade traditional missions … new operational procedures conforming to those ideas' (Rosen, 1988).
14 An innovation must change 'the manner in which military formations function in the field' in a manner that is 'significant in scope and impact' and leads to greater military effectiveness (Grissom, 2006).

15 Major Military Innovations are 'major changes in the conduct of warfare, relevant to leading military organizations, designed to increase the efficiency with which capabilities are converted to power' (Horowitz, 2010).

16 This table was reproduced from Cheung, T. M., Mahnken, T. G., and Ross, A. L. (2018). 'Assessing the State of Understanding of Defense Innovation'. *SITC Research Briefs*, Series 10 (2018-1).

17 For a concise explanation of structural realism, including offensive neo-realism see: Mearsheimer (2013).

18 Such as inter-continental ballistic missiles or stealth aircraft.

3 Research design, methodology and theoretical framework

3.1 Introduction

The underlying question for this book is how the emergence and diffusion of increasingly autonomous weapon systems would affect the stability of the Asia Pacific. However, an immediate challenge with analysing an emerging innovation such as LAWS is that prior analyses of military innovation and diffusion have generally utilised historical case studies. There are markedly fewer publications that contain projective analysis. To account for this challenge, it was important to devote space in this book to outlining how its theoretical skeleton is modified from the existing theories of military innovation, through the incorporation of elements drawn from ACT, organisational innovation, precursor wars and the Thucydides Trap. While this is a novel framework, its components remain entrenched in the established realist schools of thought, ensuring that this book remains on solid theoretical foundations.

Guided by the above theoretical framework, the research design for this book centres on qualitative, comparative case study analysis, with the primary research method being qualitative process-tracing. This research design was selected based on a review of past research in the fields of military innovation, policy diffusion and civilian disruptive innovation. Separate meta-analyses conducted by Starke (2013), Grissom (2006) and Goldman and Andres (1999), each identified that comparative case study analysis, including process-tracing, was an effective research method for analysing innovation and diffusion of policy and military technology. This book will draw upon multiple scholarly and non-scholarly sources to inform its comparative case studies, including defence whitepapers, budget statements, research fundings and official statements. This research design underpins the main contribution of this book, which demonstrates how key ASEAN member states would respond to a LAWS demonstration point and how this will impact relations of power in Southeast Asia.

3.2 Outlining the lifecycle of a disruptive military innovation

This section will provide a theoretical skeleton that will guide this book's approach to its core puzzle, understanding how the proliferation of autonomous military

DOI: 10.4324/9781003172987-3

technology to smaller states and non-state armed groups will affect the balance of power in Southeast Asia.

This framework consists of four phases that illustrate the lifecycle of a disruptive military innovation. The first phase is Foreshock: it covers the development of precursor technologies (which may in their own right be initially lauded as RMAs), their impact on the development of a disruptive weapon innovation and their proliferation once the precursor becomes normalised. The second phase is Innovation: it engages with the initial development of the revolutionary technology and the emergence of new strategic or operational doctrine that capitalises on the invention, leading to the achievement of operational praxis. The third phase, Response, begins with the demonstration point of the RMA, which triggers states to respond by adopting the technology, bandwagoning with the adopting state or developing 'balancing' alliances with other states to limit its influence. This framework departs from previous understandings of RMA diffusion at this point by acknowledging that particularly disruptive RMAs have sufficiently low adoption barriers that substantially smaller states can adopt the RMA or a derivative early in the following period of early diffusion and deployments. The fourth and final phase is Impact, which engages with the ongoing development of the initial RMA, the regional instability caused by its diffusion and the possibility of a transition of hegemonic power, at least on a regional level.

Figure 3.1 is a visual representation of the interaction between this theoretical framework, the book structure and two case studies.

3.2.1 Foreshock

There is a somewhat understandable tendency in the current discussions around AWS and AI to treat these, admittedly impactful, innovations as completely detached from the historical precedent. Putting aside the question of whether AWS are a truly revolutionary innovation in the conduct of warfare for a moment, this presumption still fails to account for the fact that in policy-making there are no truly blank canvases. As a result, even studies of disruptive innovations should account for the influence of prior similar technologies on how policymakers, military leaders and even scholars conceptualise that innovation.

In support of including this stage in the lifecycle of a disruptive military innovation, it is useful to consider the following three examples where the conceptualisation, doctrine development or deployment of previous major military advances were influenced by organisational knowledge and biases stemming from prior experiences. The first example is commonly featured in discussions of military innovations, the inter-war development of the British armoured warfare doctrine. Experiences in the First World War with the early armoured vehicles and trench warfare led the British military towards designs that focused either on infantry support roles (with trench-fording capabilities, extremely slow propulsion and a complete lack of anti-armour armament) or poorly adapted naval cruiser tactics (comparatively fast but lightly armoured vehicles focused on exploiting a breakthrough of static infantry lines). Focused on an unfinished version of the

Phases	Foreshock		Innovation		Response		Impact		
Sub-Phases	Precursor Innovation Demonstration Point	Precursor Innovation Normalises and Diffuses	Disruptive Military Technology Invented (Hardware)	Novel Operational Praxis Developed (Software)	Demonstration Point of Complete Initial RMA	State Reaction to Demonstration Point	Diffusion and Proliferation of Initial RMA	Hegemonic Competition and Power Transition	Ongoing Evolutionary Development of Complete RMA
Relevant Theories	Precursor War Adoption Capacity Theory		Military Innovation & RMA		Adoption Capacity Theory Organisation Innovation Theory		Hegemonic War Theory, Power Transition Theory, Thucydides Trap and Defensive Neo-Realism		
Case Studies	Remote Operated Unmanned Combat Vehicles		China, United States, Republic of Korea & Russia		Indonesia and Singapore		Indonesia and Singapore		
Relevant Chapter	Chapter Four		Chapter Five		Chapters Six, Seven and Eight		Chapter Nine		

Figure 3.1 Theoretical framework

innovation, the British designers did not recognise the effects of this path-dependency on their development decisions. Another example, this time from the civilian policy space, can be seen in New South Wales law, which continues to consider the internet as essentially an instantaneous version of the postal service for the purposes of making criminal threats through a carriage service (an expansion of a term that used to refer to postal or telephonic communication). In this case, policymakers capitalised on an existing well-established concept (postage service) to effectively extend the law to a new medium (the internet). As a final illustrative example, we can return to the inspiration for including precursor innovations in this analytical framework, Krepinevich's identification of the Gulf War as a 'precursor war'. Krepinevich argued that the Gulf War served as an early demonstration of a military innovation's potential in warfare (in that case information-warfare) that, while unexpected by the users and hindered by the immaturity of the underlying technology, can prompt policymakers to challenge their previous thinking and create greater impetus for experimentation.

In the case of autonomous weapons, the precursor technologies would include guided munitions, historical tele-operated platforms (such as the TT-26 Teletank), advancements in computing technologies and commercial robotics. While AI has been referred to as another precursor technology, it is far more useful to think of AI as the key enabling invention underpinning the emergence of increasingly autonomous weapon systems. It is also important to recognise that AWS is a category of emerging military technology rather than a strictly definable single weapon system or tactic. While there is certainly the potential for AWS to eventually include fully autonomous weapon platforms that replace soldiers in combat, the term would also include platforms that were given the capability to autonomous fulfil functions that they previously performed under the control or supervision of a human soldier. The potentially broad application of this term would, therefore, mean that the influence of a given precursor innovation could fluctuate based on the individual autonomous system. For the purposes of this book, however, we primarily focus on remote-operated weapon platforms especially UCAVs), because they are currently the most associated military technology with increasingly autonomous systems.

3.2.1.1 Precursor innovation demonstration point

An analysis of the potential diffusion impact of a disruptive military innovation should start with examining the evolutionary advancements that preceded their development. The deployment and impact of predecessor technologies influence the reactions of more minor states to its demonstration point.

Examining the development, deployment and diffusion process followed by the related precursor technologies provides insight into the early state reactions to the development of the disruptive technology. This is because military and civilian policymakers are generally conservative, drawing directly on the knowledge gained from the implementation of functionally similar precursor systems to inform their approach to emerging systems. This book will demonstrate how

Singapore and Indonesia (its two primary case studies) are involved in the development and deployment of the increasingly common unmanned aerial vehicle, as well as the importance of other key states that are also taking a leading role in the development of autonomous weapon platforms (including China and the United States).

3.2.1.2 Precursor innovation normalised and diffuses

The main objective of examining the development process of key precursor technologies is to identify the important decision-making processes and avenues of military innovation exchange that would influence how states would respond to the emergence of increasingly autonomous weapon systems. The principal precursor technologies of autonomous military technology are information combat systems, guided munitions and UCAVs. Using a selective case study approach, Chapter 4 of this book critically examines how ASEAN member states responded to the development and proliferation of UCAVs and related technologies.

3.2.2 Innovation

3.2.2.1 Disruptive military technology invented

Autonomous weapon systems have certainly reached the first stage of the Innovation phase; the major enabling technologies are under active development. While technology (specifically AI and machine learning) has not sufficiently developed to enable the kind of fully autonomous weapon systems that characterise the public discourse, it would be possible to design a weapon system with fully independent control over its critical functions. The crucial caveat to this is that the underlying technology has not matured sufficiently that a state would be able to deploy fully autonomous weapons on a complex battlefield without accepting a high rate of lethal error. That is not to say, however, that certain states are not actively developing weapon systems that are intended to operate autonomously (ICRC, 2016). With the rate of technological development, it is widely accepted among experts and policymakers that autonomous military technology will be sufficiently matured to enable limited deployment in a ground theatre within the next two decades. While none of the Southeast Asian states have made publicly known progress towards developing indigenous autonomous military technology, the region has become one of the largest markets for modern armaments and military technology (Dowdy et al., 2014).

To return to the example used above, UCAVs reached this stage just prior to the Balkan conflict, when the US began to use early Predators for active surveillance. The demonstration point was subsequently achieved when the Predator UCAV was armed with the Hellfire missile and utilised for a targeted, remotely operated strike. Now, more than ten years after the demonstration point of armed UAVs (Whittle, 2014), they are approaching ubiquity, with more than 80 states and multiple non-state armed groups possessing remote-operated aircraft. Singapore,

Indonesia, Malaysia, the Philippines and Vietnam have all purchased or indigenously developed UCAV platforms over the last ten years within Southeast Asia (Sayler, 2015). Their efforts to incorporate unmanned systems into their military modernisation processes have been supported by purchases of advanced systems from the United States, Russia and China (Ewers et al., 2017).

3.2.2.2 New strategic doctrine developed to capitalise on disruptive invention

The second stage of this phase is the development of a novel operational praxis (strategy, doctrine or organisational formation) that capitalises on the unique capacities of the disruptive invention. Despite the headline importance given to emerging technologies as the drivers of change, the invention of a new weapon technology is simply the first component of a disruptive military innovation. Invention must be matched with applicability to become innovation.

The history of the tank offers a good example. At the outset of the Second World War, there was no major difference in the basic armoured vehicle production technology between Germany and Great Britain. The key difference lay in the strategic doctrine that informed their development and deployment of tanks. While the German approach to tanks enabled them to capitalise on the paradigm shift brought about by armoured units (Mearsheimer, 2004), British developers, guided by a flawed approach, produced inferior tanks whose crews were trained for an outdated version of warfare. History teaches that the mere possession of technology is insufficient for a state to maintain its position through a major power transition (Silverstein, 2013); the ever-increasing rapidity of information flows in our modern globalised world has made this lesson more important than ever.

LAWS have begun to reach this stage. The ongoing international discussions at the UN Convention on Certain Conventional Weapons and the Concept of Meaningful Human Control are aspects of the process of establishing a discourse around LAWS, which will then inform the creation of competing strategic doctrines. Early attempts to establish a doctrinal approach by both states and academics (such as centaur or hybrid warfighting) have already begun (Scharre, 2016). The development of new strategic and operational concepts for the disruptive employment of a paradigm-shifting military invention is the second stage of reaching operational praxis, a fully formed initial disruptive innovation that can be demonstrated to the world.

3.2.3 Response

3.2.3.1 The demonstration point

At this point, LAWS could be viewed as initially complete. The first-mover advantage is, however, fleeting (Silverstein, 2013). Once the state uses or displays the major military innovation, other states will be forced to react to the shift in relative power stemming from its emergence. This is called the

Demonstration Point (Horowitz, 2010) and it marks the beginning of the diffusion and secondary development phase. The demonstration point is hypothesised to function differently with particularly disruptive innovations like autonomous military technology. While prior military innovations, such as the battleship or the aircraft carrier, theoretically presented these options to states upon demonstration, only major states had the resources to genuinely choose the first option, smaller states were reduced to bandwagoning together (e.g. the Cold War's Non-Aligned Movement), allying with one of the major powers or investing in other technologies to offset their loss of relative power (Silverstein, 2013). This resulted in limited competition between major powers with the resources to invest in the emerging innovation; Britain, Germany, France and Russia in the first example and Britain, the United States and Japan in the second example. The defining characteristic of a disruptive military innovation is that, at the demonstration point, it does not have sufficiently high barriers to entry to limit the number of early adopters. The remainder of this framework proceeds on this assumption.

3.2.3.2 State reaction

After any significant military innovation is demonstrated, other states will react in an attempt to preserve or improve their status within the international system. Those states with the resources, ability and organisational capacity will begin to innovate along similar lines, aiming to adopt the disruptive military innovation to protect their level of relative power. This is particularly the case when the innovation has low barriers to adoption or is more reliant upon software (knowledge, expertise, digital code, etc.) than hardware (material resources, special manufacturing process, etc.) (Grissom, 2006). For example, after the demonstration of the standoff strike capacity for UCAVs in late 2001, it was only a few months before a violent non-state actor was detained for a plot to fly a remote-operated aircraft filled with Anthrax into the British House of Commons and less than two years before Hezbollah began to utilise UAVs (provided by Iran) for surveillance (Bunker, 2015). By 2018, the use of civilian UAVs (primarily manufactured by the Chinese company DJI) by both state and non-state actors had been documented in multiple conflict zones, including Syria (Binnie, 2018), Iraq (Warrick, 2017), Afghanistan (Gramer, 2017) and eastern Ukraine (Mizokami, 2017).

However, attempting to emulate or surpass the first-mover is only one of the potential responses is only one of five options available to states confronted with a major military innovation, two internal and three external (Horowitz, 2010). States often attempt, or vacillate between, multiple responses. The two internal responses are attempting to adopt the innovation, in this case, autonomous military technology; and developing a counter-innovation. The external responses are to attempt to re-assert neutrality in the event of conflict; establish a balancing alliance against the first-mover, for example consider the formation of the Non-Aligned Movement during the Cold War; or 'band-wagon' with the first-mover state, as is the case with the American Nuclear deterrence umbrella (Figure 3.2).

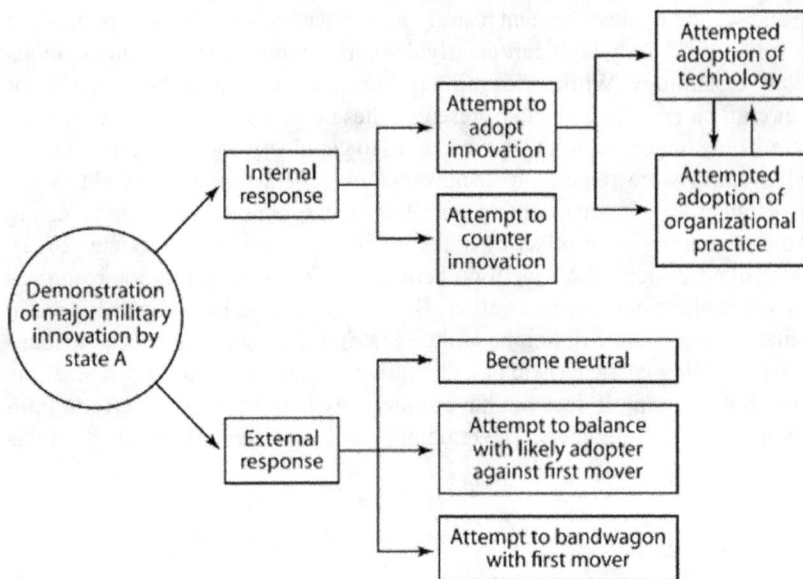

Figure 3.2 Potential state responses

This phase also sees the initial deployment of the completed disruptive military innovation, the results of which then spur further evolution. Returning to armoured warfare, the German army reviewed and altered aspects of Blitzkrieg after the surrender of France, learning from the lessons of initial tank deployments. As a more modern example, the use of UAVs has now spread to multiple states and non-state corporate entities. More than 80 states have developed the capacity to deploy UCAVs and it is estimated that every state will have reliable access to combat drones before 2025. These platforms are also getting rapidly more advanced, with some civilian UAVs now capable of greater autonomous operation than the initial block model of the MQ-9 Reaper. This secondary development is a key factor in considering the diffusion and proliferation of military technology throughout a given region.

3.2.4 *Impact*

3.2.4.1 *Regional instability*

The key impact of RMAs is that they undermine the existing paradigm of warfare. An inevitable result of this level of disruption to how states exercise power is a de-stabilisation of the international balance of power. Historically, major military innovations have enabled rising states to challenge the hegemony of more powerful states (e.g. mounted archers), for non-state actors to undermine the power of the state (e.g. firearms) or, in the case of automatic weapons, for states to impose hegemony over foreign lands. This understanding of power transition draws on the impact of disruptive innovation in the civilian business environment (Christensen, 2015).

The levelling impact of autonomous weapon systems (and increasingly autonomous military technology more broadly) contributes to the risk involved in their emergence. Due to the comparatively low entry-level adoption barriers (following a demonstration point) their emergence will have a pronounced impact on the international balance of power. Their diffusion is therefore theorised to have a de-stabilising impact on any concurrent hegemonic competition and transition, such as the one emerging between the United States and China.

3.2.4.2 Ongoing development of complete RMA

It is important to note that even disruptive innovations generally continue to develop following their initial diffusion. Some of these innovations, like cars, retain their core architecture for decades, while others might undergo continuous development or even spurts of rapid change. After their initial disruptive introduction, major military innovations have historically continued to evolve within the new warfare paradigm. Although these evolutionary changes may be dramatic, they continue to build on the paradigm established by the initial disruptive military innovation. The machine gun is a suitable example; although there have been substantial improvements made to the weapon's rate of fire, lethality and portability over the last century the basic invention at its core (gas-operated automatic cycling) is still fundamentally unchanged. Furthermore, the United States' doctrinal approach to how their infantry deploys machine guns has not changed dramatically since the Cold War. This stage in the process, therefore, reflects the ongoing improvements that will be made to autonomous weapons by their myriad of users after the deployment of the first-generation models, and the maturation and diffusion of the enabling technologies; which, based on prior innovations, are likely to centre on improving reliability, interoperability, lethality and survivability.

3.2.4.3 Transition to new balance of hegemonic power

Differing levels of mastery of the major military innovation contribute to uneven power growth within the shifting force paradigm. Any substantial shift in comparative power, particularly in a contested region such as Southeast Asia, increases the security dilemma faced by all states in the region. As elucidated in the neo-realist power transition and hegemonic war theories, while these transfers can be peaceful, historically, the emergence of paradigm-shifting weapons has generally precipitated a hegemonic war (Gilpin, 1988; Nye, 2011; Allison, 2017). This is because the initial adoption of prior major military innovations was limited to large states, ensuring their comparative advantage over minor states and limiting the scope of the conflict. The archetypical example of such a conflict was the Peloponnesian War, although other examples of hegemonic conflict (if not always armed warfare) include the Great Game, the Cold War and the current Sino-American tensions.

While the ideal situation for middle power states in a region like Southeast Asia is to exist in a stable dual hegemony, these are relatively rare and difficult to sustain. When a dual hegemonic system leads to conflict, due to the emergence of a disruptive military innovation or some other power imbalance, smaller states rarely have the security or economic capacity to alienate a potential hegemon. Therefore, it has been in the best interests of smaller states to integrate themselves into coalitions behind the competing hegemons (Ikenberry, 2016). Abiding by the existing balance of power and normative framework, gives middle power states an institutional lever to protect their interests despite their lower capacity.

Following the cessation of a hegemonic conflict, the prevailing state has historically been able to exert its influence through a favourable balance of power. The hegemonic state is then able to consolidate its position and gain resources from smaller states under its influence. This process returns the region to stability under a new power paradigm, encouraging economic growth and reducing the security dilemma of smaller states. If the hegemonic conflict does not end or, in the case of autonomous military technology, spreads intra-regional conflict, there is no guarantee that the overall balance of power would return to hegemonic stability.

This book argues that major states will not be able to maintain this comparative advantage in the case of autonomous weapon systems, de-stabilising the hegemonic competition process. Furthermore, it argues that, as more minor states gain comparative power by adopting LAWS or a derivative, regional tensions will deteriorate and the likelihood of intra-region conflict or unexpected escalation into crisis deterioration will increase.

3.3 Research design

This book draws on a composite theoretical framework to guide its exploration of the impact of LAWS as a disruptive military innovation. Its departure from the orthodoxy to examine an emerging innovation as its referent object is reflective of a key contention of this book: that LAWS have such disruptive potential that it is vital to understand what factors will influence their proliferation before they reach a demonstration point.

3.3.1 Identifying resource capacity and organisational capital capacity benchmarks

The first step in determining the most effective response for Singapore and Indonesia is to identify the level of resource intensity and organisational capital capacity required for successful adoption. Part of the challenge in determining this requirement in the case of LAWS is that a demonstration point has not yet occurred, which makes it difficult to completely eliminate uncertainty as to the final parameters of this innovation (Horowitz, 2006). This book has limited the impact of this uncertainty by carefully limiting its analytical scope to the state of enabling technology development and publicly documented operational concepts as of mid-2019. Furthermore, neither Singapore nor Indonesia possesses sufficient resource capacities to compete with the United States, China or Russia as potential first-mover states, nor would they be able to maintain a first-mover advantage in this space. Therefore, this book limits itself to considering Singapore and Indonesia as potential fast followers. Within these parameters, this book hypothesises that secondary adoption of LAWS will have a low resource capacity requirement and a medium organisational capital capacity requirement.

The increasing disparity between the resources required to procure and deploy advanced manned platforms and their remote-operated equivalents (itself a precursor innovation and an enabling technology for LAWS). One of the initial arguments in support of developing aircraft carriers was the belief that reliance on aircraft would have a lower financial requirement than battleships (Horowitz, 2010). The often-quoted 'Augustine's Laws',[1] illustrate that a similar process is occurring with modern manned platforms whose per-unit procurement and development costs continue to increase in both real terms and as a percentage of military spending. The F-35 Joint Strike Fighter is the premier example. The Australian Defence Force allocated 27.5% of its total capital expenditure in 2018–19 to the JSF, more than four times what it spent acquiring its new fleet of MHR90 multi-role helicopters (Hellyer, 2019). Traditionally, proponents have defended these cost increases with the argument that their superior combat, first-strike and survivability capabilities offset the correspondingly lower numbers that militaries could afford. However, as the pace of technology diffusion quickens (as seen with remote-operated UAVs), it will become more difficult to maintain an increasingly transient capability edge.

This incentivises militaries, particularly those of middle power states, to instead invest in procuring increasingly autonomous, AI-enabled, unmanned platforms, which have a lower resource requirement. Without human operators, these platforms do not need the same sophisticated stealth or survivability features, which would reduce the procurement and ongoing operation costs for secondary adopters. Further, unmanned platforms concentrate manpower requirements for militaries that are struggling with recruitment, shifting human soldiers away from routine, dangerous or politically sensitive roles.

Additionally, the enabling technologies for LAWS are largely dual-use in nature and have attracted significant civilian interest, investment and research, the results of which could be transferred into military platforms. Focusing here on the most important enabling technology, AI,[2] it is apparent that there would be significant overlap between the software used to enable civilian innovation and military application, for example the AI that allows a UAV to interpret LIDAR (Light Detection and Ranging) data to independently search a building for survivors following an earthquake could be used to search a building for hostile forces or civilians. There are two important caveats though. The first is that current machine learning techniques require large and task-specific datasets to 'train' a programme. For example, an encapsulated torpedo would have to be programmed with the data of potential enemy vessels prior to a conflict beginning. There is no guarantee that firms (even in the defence industry) would be able to secure this data in sufficient quantities or at the requisite specificity. The second caveat is that military platforms would require a significantly higher level of durability and 'hardening' against electronic warfare than is required in civilian platforms in order to survive in the modern battlespace. Overall, since the cost-per-unit is intentionally designed to be lower, and there is greater than expected potential for the use of dual-use enabling technologies, the resource capacity required for a secondary adopter to pursue autonomous weapon systems is hypothesised to be low, although it would be medium-high for initial developers.

Horowitz identifies three variables for measuring organisational capital capacity, Critical Task Focus, Level of Investment in Experimentation and Organisational Age (Horowitz, 2010). However, since a clear first-mover has not yet emerged this analysis cannot use the original theory's preferred benchmark. Therefore, it will draw on the diffusion of remote-operated UCVs (evaluated in Chapter 4) and the emerging evidence of experimentation by LAWS-developing states toward an operational concept for integrating autonomous military technology.

Based on prior RMAs and the characteristics of this innovation, successful integration of increasingly autonomous platforms would require Singapore and Indonesia to possess a medium level of organisational capital capacity. None of the ASEAN member state militaries has a particularly advanced organisational age, have not been involved in a major inter-state conflict in the Post-Cold War era and both the Indonesian and Singaporean militaries went through significant organisational shifts in the last 20 years. Therefore, the main points of divergence among ASEAN member states are hypothesised to be in how their Critical Task Focus affects their perception of autonomous platforms in each military domain, and their level of effective investment in experimentation over time. Reviewing the diffusion of remote-operated unmanned vehicles demonstrated that the organisational capital requirement for limited adoption is significantly lower for secondary adopters than, for example, carrier or battlefleet warfare. This view shifts, however, when one discards low-capability, unarmed aerial platforms. Only one ASEAN member state has adopted armed UAVs (Myanmar), while a handful have experimented with remote-operated ground- and maritime vehicles. Overall, therefore it is hypothesised that the organisational capital capacity required to emulate elements of a first-mover's use of LAWS will be low, but independently innovating in their operational use of LAWS would require a medium level of organisational capital capacity.

3.4 Criteria for evaluating adoption capacity

The primary purpose of this research design is to evaluate the potential diffusion pattern of increasingly autonomous weapon systems in Southeast Asia, using Singapore and Indonesia as case studies. This evaluation comprises two components: first, an assessment of the Indonesian and Singaporean adoption capacity, which then informs the second component, an evaluation of how Indonesia and Singapore respond to options based on their adoption capacity. The five response options available to states confronted with the demonstration of a disruptive military innovation were described above. As Horowitz (2010) argues, it is insufficient to merely determine which response option would be preferred by a given state, to be impactful analysis must examine which of the preferred options (or combination of responses) is most likely to succeed.

This book evaluates the capacity of Indonesia and Singapore to successfully adopt increasingly autonomous weapon systems based on five variables derived from ACT and Organisation Theory. The first variable is the state's security threat environment, the influence of traditional and non-traditional security threats on its doctrinal and procurement decisions. The second variable is resource capacity (this is referred to as 'Financial Capacity' in Horowitz, 2010) which includes military expenditure, the sophistication of the state's domestic military-industrial base and foreign arms acquisition capacity. The third variable, Organisational Capital Capacity, has three sub-variables; Critical Task Focus, Level of Investment in Experimentation and Organisational Age. The fourth variables are the receptiveness of the domestic audience towards autonomous military technology, and the final variable is the military's demonstrated capacity to develop or emulate the specialised operational praxis required to effectively deploy the disruptive invention. The following section will explain each of these variables in greater detail.

3.4.1 Security threat environment

The first adoption variable is the extent to which the primary traditional and non-traditional threats to a state affect its defence policymaking and expenditure. The first important factor to consider when applying this variable is the historic tensions between Southeast Asian states. The core of this variable is identifying the key significant security threats in the minds of strategic planners in these states, which then has an influence on how resources are committed. Through reviewing government working papers, such as the 2017–19 issues of the *Singapore Terrorism Threat Assessment Report* and scholarly papers from authors such as Syailendra (2017) and Santikajaya (2016), this book identifies territorial intrusion, terrorism and piracy as regional security threats that would influence how Indonesia and Singapore perceive the value of autonomous systems. It draws on the broader existing literature from authors such as Ray (2018), Hoesslin (2016), Haacke (2009) and Purbrick (2018) to inform its analysis of how Indonesia and Singapore are responding to their evolving threat environment.

3.4.2 Resource capacity

Evaluating whether Indonesia and Singapore possess sufficient resources to adopt the increasingly autonomous military technology requires analysis beyond a simple budgetary analysis to a more holistic consideration of the state's capacity to direct its economic, technological and political resources; therefore, government white papers, defence-spending disclosures and doctrinal documents are all important sources. Both Indonesia and Singapore irregularly publish defence white papers, and both militaries maintain in-house research journals, which proved to be a useful source of doctrinal information. Unfortunately, while Indonesia publishes some military-spending data as part of the annual national budgetary papers and allowed the United Nations to publish historic spending figures, official data on Singapore's military spending from official sources was less accessible and had to be supplemented.

To supplement the insufficient availability of official budgetary data, this analysis draws on non-government research published between 2014 and 2019, a timeframe that ensured relevance. These alternative sources included scholarly literature published during this period from authors including Laksmana, Bitzinger and Raska. In addition to traditional literature, this book draws on think tank publications, such as the 2018, 2019 and 2020 issues of *The Military Balance*, published by the International Institute for Strategic Studies and reviews of military expenditure published by the Stockholm International Peace Research Institute during the same period. This section also draws on the 2018 *Defence Economic Trends in the Asia Pacific* report, which is published annually by the Australian Department of Defence as an official reference guide.

Corporate defence industry research proved vital in evaluating this variable, particularly in the case of Singapore. Sources include industry outlook reports published by the McKinsey Institute (Dowdy et al., 2014) and the 2015 *Deloitte Asia-Pacific Defence Outlook* report (Bars, 2015). Finally, this analysis draws on market research from IHS Janes, from which the author accessed data on Singaporean military spending and specific research allocations for the Indonesian military. Drawing on a broad selection of documents supported a more effective analysis of the resource capacity of these states than could be derived from the limited official military spending data.

3.4.3 Organisational capital capacity

The second adoption variable is the state's organisation capital capacity, which 'represents a virtual stockpile of change assets needed to respond to changes in the character of warfare', and has three sub-variables: Critical Task Focus, Level of Investment in Experimentation and Organisational Age (Horowitz, 2010). In addition to the previously mentioned sources, assessing this variable draws on research published by Matthews and Yan (2017), Sebastian and Gindarsah (2013) and Arif and Kurniawan (2018), as well as working papers from researchers at the S. Rajaratnam School of International Studies. Addressing this variable also draws on the official English translation of the 2015 Indonesian Defence White Paper (the most recent iteration), the 2019 *Singapore Terrorism Threat Assessment Report* and speeches made in 2017 and 2018 by Singaporean government officials for evidence of Critical Task Focus. This variable also utilises data drawn from a combination of media articles, government press releases, the official statements of the Non-Aligned Movement and publications in Indonesian and Singaporean military journals written by serving (or retired) military personnel, which are reflective of the emerging nature of state responses to autonomous weapon systems.

3.4.4 Receptiveness of domestic environment to innovation

Unfortunately, when evaluating Indonesia and Singapore against this variable, there is a dearth of specifically applicable quantitative data published on public opinion towards artificial intelligence in a military context or autonomous systems, which is a major gap in the literature that should be the subject of future inquiry. Early studies on public opinion towards LAWS include studies authored by Carpenter, Open Robo-Ethics Institute and Horowitz, which were all primarily focused on the United States. More recently, the Campaign to Stop Killer Robots commissioned two studies (2017 and 2019); however, these did not directly survey citizens of Indonesia or Singapore.

Without direct data, this book draws on published research that examined the Indonesian and Singaporean citizens' attitudes towards the United States' drone strike programme, as well as relevant government legislation, regulation of civilian use of remote-operated UAVs in these states and submissions to the Centre for a New American Security's Proliferated Drones report series (Desker and Bitzinger, 2016; Rahakundini and Prasetia, 2016). This is further supplemented by reviewing official statements from government ministers and the Non-Aligned Movement, evidence of corporate investment in modernising defence industry capability toward autonomous platforms, research investments in this technology and position statements from leading universities, for example, the Indonesian *Universitas Gadjah Mada* joined the Campaign to Stop Killer Robots in November 2018. Taken together, these alternative sources provide sufficient data to inform a position on whether the public would effectively oppose the introduction of military platforms with increasing levels of autonomous capability.

3.4.5 Ability to develop or emulate a specialised operational praxis

The final variable is the ability of the state to develop or emulate a specialised operational praxis. Similar to the previous variables, applying this variable to the case study states draws on evidence from scholarly literature published during the 2014–19 period. Primary authors utilised for this variable include Laksmana, Raska, Rosin and Bitzinger. Academic sources were then supplemented by working papers published by IHS Janes and the *Mapping the Development of Autonomy in Weapon Systems* report, published by the Stockholm International Peace Research Institute. This is complemented by the accounts and analysis of prior major military innovations contained in existing scholarly diffusion literature.

This variable also draws on defence white papers, graduate papers written by serving military officers from Indonesia, Singapore and the United States and working papers published by the Indonesian Ministry of Defence and Singaporean Defence Science and Technology Agency. This combination of resources provides evidence crucial for process-tracing the diffusion of operational praxis of prior military innovations, which offers an insight into how these states would respond to the emergence of autonomous weapons. An understanding of the process by which Singapore and Indonesia are able to develop or emulate a novel

Table 3.1 State adoption capacity variables

Adoption Variables
Security Threat Environment   Resource Capacity   • Military Expenditure   • Capacity of Domestic Military-Industrial Base   • Foreign Arms Acquisition Capacity   Organisational Capital Capacity   • Critical Task Focus   • Level of Investment in Experimentation   • Organisational Age   Ability to Develop or Emulate A Specialised Operational Praxis   Receptiveness of the Domestic Environment to Innovation

operational praxis for the deployment of autonomous military technology is highly useful because states are intelligent actors that are influenced by the public actions of larger neighbours and first adopters (Table 3.1).

3.5 Regional focus and case study selection

The primary method for this book is a qualitative, cross-case comparison based on the five adoption capacity variables set out below. This is supported by documentary analysis and policy process-tracing. This research design is particularly suited for studying multiple state diffusion of technology (which is rarely statistically quantifiable while ongoing) because it will also include non-scholarly budgetary, policy, technical and doctrinal evidence alongside the existing academic writings. Indeed, this variety of selective case study approaches has been utilised prolifically across scholarly research, government policy papers and military doctrinal notes and was explicitly advocated by Starke (2013) in his evaluation of methods for researching policy diffusion.

Due to space and resource limitations, it was necessary to limit the analysis in this book to centre principally around two case studies, which in turn needed to be linked in some manner to allow for comparison and analysis of strategic impact. Given the strategic focus of this book and the comparative nature of its core analytical tool, it was sensible to choose case studies that were geographically linked and had access to comparable resource capacities. When determining the geographic focus for this book, I primarily considered three regions; Southeast Asia, Sub-Saharan Africa and Northeast Asia. Each of these regions has geostrategic significance to the dominant great power states who are engaged (directly and indirectly) in the development of AI and Autonomous Systems. A Northeast

Asian focus would have led to the book centring on the impact of autonomous weapon system proliferation on the stability of the Korean Peninsula, a key geostrategic flashpoint in the emerging great power competition between China and the United States. Equally, however, both China and the United States have increasingly been investing diplomatic and economic resources in expanding their influence in Sub-Saharan Africa and Southeast Asia, reflecting the importance of these regions to their hegemonic competition. These regions also have the benefit of hosting multiple smaller and middle power states that have their own intra-regional political tensions as well as a number of active violent non-state groups.

I chose to focus on Southeast Asia over Sub-Saharan Africa principally because the increasingly tense South China Sea territorial disputes and the recent increases in military spending among the strongly nationalistic states in the region suggested that its importance as a geostrategic flashpoint would increase. Furthermore, Southeast Asian states were not represented in the international discourse around autonomous weapon systems and the literature remained curiously silent on the potential impact of this technology in the region, despite strong military modernisation efforts and a demonstrated interest in the acquisition and deployment of remote-operated drones.

This book uses Indonesia and Singapore as its primary case studies to evaluate the potential impact of the diffusion of autonomous weapon systems on regional security and stability in Southeast Asia. These states share a common regional location and are founding member states of the ASEAN, which reflects the broader analytical focus of this book on Southeast Asia. This focus reflects the fact that Southeast Asia is among the fastest-growing and politically influential geographic regions, especially in light of the growing hegemonic tension between China and the United States. As part of a wider shift in regional power across the Indo-Pacific, ASEAN member states are predicted to grow in influence and relative power. While they have generally profited from the international stability maintained by the 'international rules-based order', one of the foundational purposes of ASEAN was to discourage attempts by great power states outside the region to dictate policy within Southeast Asia (Kuik, 2016). Unlike during the Cold War, as we move toward this hegemonic conflict some Southeast Asian states have undergone significant economic growth, which has been translated into a noted regional military modernisation effort that has turned Southeast Asia fastest-growing arms importation market globally (Dowdy et al., 2014); although neither increase has been uniform across ASEAN membership. This confluence of factors makes Southeast Asia a more effective source to draw case studies from than East Asia or the Middle East, despite its lack of states that are designing autonomous military technology.

It is in this environment that increasingly autonomous weapon systems are emerging. The decision to focus this book on Indonesia and Singapore is reflective of their influence within Southeast Asia, the variety of active ongoing traditional and non-traditional threats to regional security and their status as the greatest military spenders among ASEAN member states.

3.5.1 Case study 1: Indonesia

The first case study focuses on Indonesia, a leading economic and military power within Southeast Asia. Indonesia exercises significant influence among the ASEAN member states and has already begun to hedge in its attachments to China and the United States in response to rising tensions between the two great powers. Situated at the southern edge of Southeast Asia, Indonesia is the world's largest archipelagic state, spread across a wide spread of thousands of islands (from Aceh to Papua); while some are uninhabited, others are densely populated. For example, the current capital, Jakarta, is situated on Java, which is the most populated island on the planet. Indonesia is the seventh-largest state globally in terms of combined land-sea territory and is home to over 261 million people. The ethnic makeup of this population is diverse, while it hosts the largest Muslim population in the world, the Indonesian population speaks more than 300 native languages (BBC Asia, 2018). Although there is a strong element of nationalism running through Indonesian society (Wijaya, 2018), this is not simply a natural element of Indonesian culture; rather, it was deliberately stimulated during the post-Suharto transition to democracy. The recent presidential election illustrated the illiberal turn that Indonesian democracy has taken, with long-standing clientelist practices and concerns about inequality fuelling and, increasingly religiously influenced, nationalist undercurrent that is in turn being courted by political leaders (Diprose, McRae and Hadiz, 2019). Over this system looms the Indonesian military which, despite having no official political role and stridently proclaiming its neutrality, is widely considered to remain an important variable in domestic politics (Laksmana, 2019).

After being among the hardest-hit countries by the 1997 Asian Financial Crisis, Indonesia has now emerged as the largest economy in Southeast Asia, with its GDP rising from USD 95.45 billion to USD 1.04 trillion in the 20 years to 2018 (The World Bank, 2019). However, this economic growth has not been equally distributed, on a per capita basis Indonesia's GDP is only ranked tenth in Southeast Asia as of 2019 (based on IMF estimate) (International Monetary Fund, 2019) and inequality remains high with around 10% of Indonesians living below the poverty line (The World Bank, 2017). Rather than improving internal welfare, a significant portion of Indonesia's recent wealth has been committed to the ongoing efforts to modernise the Indonesian military, presumably in response to rising regional tensions. An important aspect of this modernisation has been a renewed emphasis on developing a globally competitive domestic defence production capability. Beyond a level of self-sufficiency and resilience from sanctions (an enduring concern for the Indonesian military), this offers economic benefits and improves a state's soft power influence over its customers. There is also a level of nationalistic prestige to be gained through producing and exporting arms. Pursuit of this prestige was a driving factor behind the Suharto regime's development of state-owned arms producers in the 1970s (Bitzinger, 2017).

Indonesia is a valuable case study because it is a rising middle power Southeast Asian state and an influential member of ASEAN that has demonstrated an interest in unmanned systems and networked warfare. Indonesia is faced with multiple non-traditional security threats (such as piracy, climate change, persistent poverty and terrorism) against the backdrop of an encroaching China and simmering intra-regional tensions with its neighbours. Furthermore, Indonesia has demonstrated a past willingness to militarily intervene in territory that is nominally independent or under the influence of other states (such as East Timor). Finally, Indonesia maintains the second-largest military budget among ASEAN member states and maintains ongoing arms exchange relationships with key developers of autonomous military technology.

Based on its economic and military growth, Indonesia's influence within Southeast Asia and beyond is expected to grow substantially over the next two decades. While this growth is not unique to Indonesia among ASEAN member states, the combination of this economic growth and military modernisation efforts within the context of Indonesia's myriad of security threats and unstable provincial politics results in a unique case study for understanding the impact of state and non-state actors adopting weapon systems of increasing autonomy or their derivatives.

3.5.2 Case study 2: Singapore

The second case study state utilised by this book is Singapore, a fellow founding member of ASEAN. Singapore is a major economic centre that is highly dependent on international trade and commerce and has close security ties with the United States, which reflects Singapore's strong foreign policy goal of contributing to regional stability. While it maintains the most advanced military among ASEAN member states, Singapore remains in a precarious security position. Singapore is surrounded by rival states, borders the South China Sea territories (although is a non-claimant state) (Chan, 2016) and is reliant on secure international commerce, a reliance that emphasises the continuing threat of piracy and violent non-state actors in the region. Singapore has significant severe geographic constraints and is beginning to feel the effects of ageing on its, comparatively, a population (Jamrisko and Amin, 2017). These factors make increasingly autonomous military technology, and fully autonomous weapon systems, highly attractive to the Singaporean military. Indeed, unmanned platforms are central to the Next Generation Singapore Armed Forces strategic concept published in 2019 (Wong, 2019), and have declared their interest in pursuing artificial intelligence for military purposes.

Examining Singapore as the second comparative case study for this book presents a decidedly different Southeast Asian perspective than Indonesia while retaining the geographic focus of this book's analysis. Singapore is an established power in the sub-region operating under a British-derived, albeit utilitarian and somewhat authoritarian, democratic system. Singapore's GDP is 13 times higher than the Southeast Asian average in per capita terms, however, its economic strength is highly reliant on international trade. Given that Singaporean policy-makers have consistently stated that Singapore's capacity to maintain a regional economic and security edge is dependent on advanced technology, it is unsurprising that this city-state has made independent inroads into the development of key enabling technologies for autonomous weapon systems.

While both states have identified violent non-state actors as major security challenges and have a vested interest in maintaining regional stability and a consistent balance of power, their motivations and security focuses are different. Furthermore, while both states maintain strong defence ties to states with autonomous military technology, the nature of these relationships is quite different. Singapore has been matching its strong history of emulating the military practices of larger states with major recent military purchases from the United States, while Indonesia has a far more diverse arms supplier base, remaining wary of the threat of a renewed arms embargo. Finally, Singapore's involvement in the primary threat to regional stability, the rise of China and the connected territorial disputes is demonstratively different from that of Indonesia and other ASEAN states. Combined, these factors mean that choosing Singapore as a case study will provide a unique perspective on the role of autonomous military technologies in future regional security efforts.

3.6 Limitations

Despite its comparatively recent rise to scholarly prominence the emergence of autonomous military technology and its impact on the conduct of warfare is a significant topic of inquiry and, as with any scholarly work, this book must be limited in its ambition and scope. The aim of this book is to critically analyse how Southeast Asian state responses to the emergence of Lethal Autonomous Weapon Systems will influence its impact on the balance of power in South-East Asia. To preserve space and resources in pursuit of its aim, this book must curtail its exploration in three key areas. This book does not engage critically with the scholarly debate over what qualifies an innovation as a Revolution in Military Affairs, its analytical focus is geographically limited to South-East Asia and it contains a limited analysis of domestic level policy decisions other than where directly relevant to autonomous military technology. While some of these limitations have been addressed in other scholarly works, others offer an opportunity for further research in the future.

The first limitation is that this book does not contain a detailed discussion of what qualifies a given military innovation as an RMA and whether LAWS qualify under each of the myriad, competing approaches debated in the existing literature. Within the broader military innovation scholarly community, the concept of RMA has gone through stages of criticism and rigorous debate. Given that this debate and the greater innovation literature is discussed in great detail in the literature review section, it is sufficient to state here that there is an ongoing and fierce debate over what characteristics define an RMA in comparison to, for example, a major military innovation or a disruptive innovation (in the commercial sense). This debate generally centres on whether specific innovations (such as the stirrup) rise to the level of an RMA (Vickers, 2010), a discontinuity in the paradigm of conflict. This book makes the assumption that the development of weapon systems capable of fully autonomous operation, entirely removing the human from the immediate decision to end life in a combat zone, would comprise a discontinuity. The author accepts that other scholars may disagree with this assumption and aims to contribute in separate research efforts to the debate on whether a LAWS should be considered an RMA.

The second limitation of this book is its narrow geographic focus on Southeast Asia. Southeast Asia was chosen as the main geographic focus of this book for three reasons. Firstly, when considering the emerging hegemonic conflict between China and the United States, ASEAN states are the nearest and most influential collection of middle power states. Secondly, Southeast Asia a region of immense economic and geopolitical importance yet is riven by regional tensions. ASEAN member states maintain historic rivalries and mutual distrust, which have not been helped by the emergence of China as a competing hegemonic influence. It is this rivalry and security dilemma that is partially blamed for the drastic rise in military spending by ASEAN member states over the past decade (Simon, 2012; Dowdy et al., 2014; Fleurant et al., 2017). Finally, Southeast Asia hosts numerous violent non-state actors that directly impact ASEAN member state security and will affect how they engage with autonomous military technology. In addition to long-running insurgencies (for example, in the Philippines and Indonesia) ASEAN states are struggling against the influence of transnational criminal groups, which are widespread and influential in a region that is a well-known hub for piracy, human trafficking, drug trafficking and the illegal flora and fauna trade. In 2016 the United Nations Office on Drugs and Crime estimate the annual value of organised crime in the region at US$100 billion (2016). For the above reasons, Southeast Asia was chosen as the geographic focus of this book.

However, this limitation has four analytical impacts that reflect this narrow geographic focus and are worth highlighting. Firstly, this book does not devote meaningful analysis to the role of Russia, despite its significant role on the international stage and the fact that it has admitted to active development efforts of autonomous military technology. A deeper analysis of Russia's involvement was omitted because it has a lesser role in this sub-region compared to the United States and China. Further exploring the influence of Russia in the initial diffusion of autonomous military technology is another interesting path for future research. Additionally, this book does not attempt to review comprehensively or analyse Australian or American defence policies or procurement patterns; to do so is beyond the scope of this book's underlying puzzle. Along similar lines, this book is not an economic analysis of power or a comprehensive analysis of the influence of Chinese economic hegemony in the region. Conducting a detailed analysis of the impact of international trade routes and Chinese domestic economic policy is beyond the scope of this book. Those who are interested in the growing influence of Chinese economic hegemony in the region are recommended to review Allison (2017) or Ikenberry (2016).

Finally, this book does not meaningfully engage or critically analyse domestic level policy except to the extent that it is relevant to analysing the regional impact of proliferated autonomous weapons. The impact of autonomous technology on purely domestic problems such as privacy, domestic state surveillance, civil rights and law enforcement use of force policies, is beyond its scope. For more practical reasons directly stemming from time and resource constraints, this book does not attempt to address the gap in the scholarly literature about public opinion toward autonomous weapon systems and the factors that influence this opinion. Horowitz (2016) and the Open Robo-Ethics Initiative (2015) have very interesting articles available on this topic. However, these articles are focused largely on the United States, leaving a prime research opportunity to engage with public opinion among ASEAN member states.

3.7 Conclusion

While recognising that actors in an innovation process often attempt, or vacillate between, multiple responses options, the structure of this research design reflects the diffusion cycle of military innovation through the lens of its novel theoretical framework. This framework provides an analytical structure to guide the application of the research methods that underpin the book's main contribution, which is evaluating the capacity of Singapore and Indonesia as case study regional middle powers to respond to Lethal Autonomous Weapon Systems as an emerging disruptive military innovation, and the impact this will have on security and stability in Southeast Asia.

The core methodology of this research design is a case-study-based approach, supported by process-tracing and documentary analysis of identified adoption variables. This is particularly suited for studying military diffusion and has been utilised by a number of scholars in the fields of policy diffusion, military innovation and disruptive commercial innovation. Alongside traditional academic literature, this research draws extensively on a combination of data and analysis from defence research bodies, civilian state agencies and non-government think tanks.

The structure of this book reflects the four phases that engage with the lifecycle of a particularly disruptive major military innovation or rapidly proliferating disruptive military innovation. The first phase is Foreshock: it covers the development of precursor technologies (which may in their own right be initially lauded as RMAs), their impact on the development of a disruptive weapon innovation and their proliferation once the precursor becomes normalised. In the case of LAWS, the precursor technology was Unmanned Combat Vehicles, principally armed Unmanned Aerial Vehicles, whose demonstration point occurred in 2001 with their use by the United States as a strike weapon. Chapter Four explores how Southeast Asian states have interacted with and been impacted by the diffusion of armed UAVs.

The second phase is Innovation; it engages with the initial development of the revolutionary technology and the emergence of new strategic or operational doctrine that capitalises on the invention, leading to the achievement of operational praxis. Applying this phase to autonomous weapon systems occurs in Chapter

Five, which evaluates the current development of the hardware and software components of LAWS by the key developing actors (such as the United States and China).

The third stage, Response, begins with the demonstration point of the disruptive military innovation, which triggers states to respond to the shift in the balance of power. This relates directly to the core research questions of this book and, as such, is the main application of the two case study states. Chapters 6 and 7 apply the adoption capacity variables to Indonesia and Singapore, while Chapter 8 addresses how adoption would factor into the responses of these states to a future LAWS demonstration point.

The final stage of this framework is Impact. This phase covers how the international community goes about the ongoing development of the initial disruptive military innovation, the regional instability caused by its diffusion and the possibility of a transition of hegemonic power, at least on a regional level. This stage is reflected in the latter sections of the book, which evaluates the impact of LAWS proliferation in Southeast Asia and the hegemonic transition conflict between the United States and China

Notes

1 'In the year 2054, the entire defense budget will purchase just one aircraft. This aircraft will have to be shared by the Air Force and Navy 3 1/2 days each per week, except for leap year, when it will be made available to the Marine Corps for the extra day' Fallows, 2002
2 AI is a broad term, in this instance I am utilising Horowitz's definition: Artificial Intelligence can be described as 'the use of computing power, in the form of algorithms, to conduct tasks that previously required human intelligence' Horowitz, 2019.

4 Development and diffusion of unmanned combat vehicles

4.1 Introduction

While the emergence of lethal autonomous weapon systems is the focus of this book, military and civilian policymakers have consistently drawn a conceptual link in their public commentary to the existing remote-operated platforms, primarily unmanned aerial vehicles. This practice reflects the fact that policymakers are generally conservative and draw directly on knowledge gained from the implementation of functionally similar precursor systems to inform their approach to emerging innovations (Goldman and Andres, 1999). Therefore, this chapter is comparative in nature, offering an analysis of how Indonesia and Singapore are responding to the proliferation of remote-operated unmanned combat vehicles (UCVs). Its purpose is to identify factors, institutions and actors within these militaries that would inform how Indonesia and Singapore would conceptualise increasingly autonomous weapon systems.

Examining prior major military innovations demonstrates this tendency for prior experience with the development, deployment and diffusion cycle of precursor technologies inform how states respond to emerging technology. A prime example of this effect can be seen in inter-war British tank doctrine and design philosophy, which drew on their experiences in the First World War and naval warfare, splitting their efforts between the infantry tank and the cruiser tank, with neither proving well-suited to the armoured warfare paradigm pioneered by their German counterparts, which instead focused efforts on unit-level radio communications, combined arms operations and integrated aerial support. Therefore, analysis of an emerging disruptive military innovation should begin by examining the development and diffusion of its precursor innovation. In the case of LAWS, a significant precursor innovation is UCVs, which are distinguished by the fact that their 'critical functions' remain under the control of a human operator, albeit remotely.

The first section of this chapter outlines the current status of unmanned combat vehicles across the aerial, maritime and terrestrial combat domains utilising prominent examples and explains the emerging operational concepts underpinning their ongoing development. This is followed by an examination of the role of key states in the development and proliferation of UCVs. At the core of this

DOI: 10.4324/9781003172987-4

chapter is the evaluation of Indonesia and Singapore against the adoption variables (identified in Chapter 3) using data from the years following the initial demonstration point of armed UAVs. This chapter will also argue that factors that contributed to the evident preference among ASEAN member states for low-cost, unarmed platforms over complex strategic strike-capable variants would inform their preferred response in the event of a future LAWS demonstration point.

Overall, this chapter demonstrates that rising resource capacities in the region and the centrality of a dual-use technological component (remote-operated aircraft) contributed to the successful adoption of remote-operated UAVs by leading ASEAN member states. However, it also establishes that security environments, critical task focus and regional tensions had a greater impact on the success of these attempts than comparative adoption cost. This should offer pause to those who argue that only large, advanced states will be important actors in the early post-demonstration period of lethal autonomous weapon systems.

4.2 The current status of unmanned combat vehicles

The precursor innovation for increasingly autonomous weapon systems, remote-operated unmanned combat vehicles, has proliferated at a remarkable rate. As with any innovation, understanding this proliferation requires an exploration of the development of its technological (hardware) and organisational change (software) components. The purpose of this section is to demonstrate how the ongoing development of remote-operated weapon platforms is mutually influencing the emergence of a series of identifiable praxes which would, in turn, influence the policymakers' approach to increasingly autonomous weapon platforms.

The most prominent form of UCV has been unmanned aerial vehicles (also known as 'Drones'), which have also been the primary subject of media coverage and are therefore the version of military robotics foremost in the mind of the general public. However, this should not be allowed to minimise the military importance of unmanned ground and maritime vehicles, which are also perceived as critical to the security of ASEAN member states by defence policymakers. While the past 15 years have only seen five actors use armed UAVs (drones) in combat (United States, United Kingdom, Hezbollah, Israel and Pakistan); seven other states possess deployable armed drones, 10 are developing combat drones, and 50 (including Australia) are developing domestic production capability. There are 80 states that have acquired some form of combat deployable drone technology, and it is estimated that every state will have reliable access to combat drones before 2025 (Sayler, 2015), with global expenditure expected to reach US$91 billion by 2024 (Stohl, 2015).

4.2.1 Unmanned aerial vehicles

Given their status as the most prominent and well-funded variety of unmanned combat vehicle, it is best to start with UAVs. The recent wave of interest in remote-operated combat aircraft arguably began with the development of the

MQ-1 Predator and its deployment by the United States in the Balkan conflict. Although its demonstration point was arguably the first use of an armed UAV in a stand-off strike role, which occurred in late 2001, development of the technology and doctrine aspects continued. While this was not the first modern UAV to be used by the US military, the armed Predator captured the public fascination and its image (and that of its larger successors) became synonymous with unmanned weapon systems.

UAVs are typically divided by endurance and flight altitude; for example, the MQ-9 Reaper is considered a Medium Altitude, Long Endurance (MALE) UAV. However, because there is no universal agreement as to what specific benchmarks should be used, a definitional grey area remains. To avoid confusion, this book includes the categorisation system published in the US Army's Unmanned Aircraft Systems Roadmap 2010–35 (Figure 4.1).

While the majority of the 80 states that possess UAVs only have access to unarmed Intelligence, Surveillance, Reconnaissance (ISR) models, this is changing; ten states (including Iraq, Myanmar, Pakistan and Turkmenistan) had acquired armed UAVs from China by the end of 2017.[1] While the US UAVs are generally significantly more capable than their competitors, due to their advanced technology and sophisticated information infrastructure, they proved significantly more difficult to acquire. This was largely due to a previous policy position that explicitly restricted the sale of armed UAVs (including those equipped with a laser designator); however, adherence to the Missile Technology Control Regime further limited the export of UAVs with a payload above 500 kg. As a result, only a small number of the US allies have successfully purchased US UAVs, including Australia and France.

Although UAVs offer substantial benefits over manned aircrafts, including reduced economic cost, reduced risk to personnel and longer mission endurance, they are also far more vulnerable to conventional air defence methods and electronic warfare, which limits their impact. Additionally, states that are unaligned with the United States are limited by the technological and data processing obligations of comparable long-term UAV surveillance. The US Air Force found that 83 personnel were required to process the information gathered during a single operational flight by an MQ-9 Reaper. By 2012, they had accumulated a processing backlog of over 400,000 hours of footage (Gregory, 2011). The United States

UAS Category	Max Gross Take-off Weight	Normal Operating Altitude (Ft)	Airspeed	US Army UAS Example
Group 1	<20 pounds	< 1200 Above Ground Level	< 100 knots	RQ-11B Raven
Group 2	21 - 55 pounds	< 3500 Above Ground Level	< 250 knots	No Current System
Group 3	<1320 pounds	< 18,000 mean sea		RQ-7B Shadow
Group 4	> 1320 pounds	level (MSL)	Any Airspeed	MQ-1C Gray Eagle
Group 5		> 18,000 mean sea level (MSL)		No Current System

Figure 4.1 US Army UAS MTOW classification

is the only state that currently has the data processing and linkage capacity to operate UAVs on an intercontinental scale. While only a handful of states have used UAVs in lethal operations to date, the underlying technology is rapidly proliferating globally. China, Russia and Iran have domestic production capacity, and even minor powers such as Nigeria, Pakistan and Iraq have acquired armed UAVs (Ewers et al., 2017). While they are not comparable to their US counterparts, they are still relatively effective platforms. Furthermore, with the exception of Brunei (Singapore Ministry of Defence, 2018), there has been a clear resource commitment among ASEAN states to improve their capability to produce or purchase military UAVs.

Finally, there has been a boom in civilian manufacturing and the sale of commercial unmanned aircraft. While no serious comparison can be made with military models, Commercial Off the Shelf (COTS) drones are becoming ever more advanced, a factor that has already contributed to their use by violent non-state actors. This boom has direct relevance to increasingly autonomous weapon systems, which also rely on dual-use technologies (such as machine learning-based artificial intelligence algorithms, computer vision and sensors).

4.2.2 Unmanned ground vehicles

An unmanned ground vehicle (UGV) is a platform that operates in contact with the ground without the physical presence of a human operator. Typically controlled remotely, they are designed to extend the capabilities of human soldiers or to undertake 'dirty, dull, or dangerous' roles under supervision, rather than to operate independently. Remote-operated ground platforms are appealing to states with inaccessible or remote land borders, which are difficult to police, monitor and defend, particularly in the case of internal unrest. Myanmar's northern border region is a good example of this (Jenne, 2017). In other cases, UGVs could be suitable to monitor contested or disputed land borders, which otherwise generate tension or even sporadic conflict, such as between Thailand and Cambodia (Kocak, 2013), and India and China's Himalayan border (Medcalf, 2014). The importance of understanding how states are approaching unmanned ground vehicles is highlighted by the existence of the Super-Aegis II, a supervised autonomous weapon platform which has already seen limited deployment by South Korea.

Existing UGVs can be divided along task lines. Firstly, there are unmanned ground combat vehicles, which are typically large and heavily armoured. Resembling a tank, the role of these vehicles is to participate directly in combat while reducing risk to human soldiers. These platforms are operated remotely by other human soldiers who are nearby but not (typically) in direct contact. Modern examples include the SWORDS platform (United States), the Sharp Claw (China) and the Uran-9 (Russia). In May 2018, the latter became the first armed UGV (reportedly with the capability to operate autonomously) to be deployed directly into modern combat zone. This variety of direct combat UGV appears to have appealed to Russian defence planners, who have several operationally similar vehicles under development.

The second type of unmanned ground vehicles are those designed for explosive disposal. They are generally operated from a short distance away with simple controls. Examples of this variety of UGV include the Packbot and the MARCbot. The lethal use of this kind of UGV was dramatically brought into the public eye in July 2016 when police in Dallas, Texas, used a similar UGV (Remotec Andros Mark V-A1) carrying a C-4 charge, to kill an armed suspect during an active shooter incident. Prior to this, law enforcement agencies had used Packbot style UGVs for a variety of non-lethal tasks beyond their intended bomb disposal duties, from surveillance to removing a blanket to see if a suicidal individual was armed. Whether the Dallas police were entitled to utilise a UGV in this way is beyond the scope of this book, but it did set an interesting precedent regarding the use of unmanned platforms for law enforcement and the sub-national exercise of state power.

The third variety of unmanned ground vehicles are designed as re-supply and logistics tools. They are typically lightly armed or unarmed and either are remotely operated or possess a limited, task-based autonomy that allows them to follow a command signal carried by friendly soldiers. Prominent examples of this type of UGV include the Big Dog and Alpha Dog (which were both cancelled) as well as the Israeli RoBattle and the US Crusher. The final variety are armed, fast vehicles that are capable of limited task-based autonomy but are generally remotely operated. These are typically intended for defensive patrols and can remotely engage intruders. Examples include the MDARS-E and the Guardium, which are both capable of lethal force.

Land-based operation presents the most complex environment for any level of autonomy, especially in the context of low-intensity or irregular conflicts. This makes it impossible to deploy current generation autonomous technology on ground-based weapons without an unacceptable level of risk to civilians and friendly combatants. It is worth noting that this is not as severe a restriction in remote or inaccessible border areas where there is only a small concentration of civilians and non-legitimate targets in the event of a conflict. Overall, it is far more likely that, at least in the near future, remote operation of unmanned ground vehicles will remain the norm, especially among smaller militaries.

4.2.3 *Unmanned maritime vehicles*

The final category of unmanned combat vehicles is unmanned maritime vehicles (UMVs). This category includes both unmanned surface vessels (USV) and unmanned underwater vehicles (UUV), both of which have been employed in a wide range of military and civilian uses. Given the importance of maritime boundaries and controlling violent non-state armed groups within the Southeast Asian security environment, it is unsurprising that ASEAN member states have been closely following the development of UMVs. The emerging operational praxes around their remote-operated predecessors would indicate that the use of increasingly autonomous platforms in Southeast Asian waters is likely to continue to revolve around surveillance, force protection and area denial.

The maritime environment is generally considered the least technically and ethically challenging combat theatre for deploying autonomous and remote-operated weapons (Boulanin and Vebruggen, 2017). This is a comparative statement, as there is still risk involved in deploying unmanned platforms into a region that is characterised by ongoing territorial disputes, inter-state tensions and multiple armed non-state actors. For example, consider the December 2016 interception and seizure of a US Navy unmanned underwater vehicle in the Philippines Economic Exclusion Zone by a People's Liberation Army Navy vessel (Perlez and Rosenberg, 2016). The UUV was returned less than a week later following a formal complaint by the US military (Lin-Greenberg, 2016). This incident highlighted one of the key risks posed by deploying UMVs in Southeast Asia: that a state could react unexpectedly towards an unmanned platform, potentially due to differing interpretations of their status under international law, a scenario that is even more volatile when the UMV is operating in disputed maritime territory.

A major cause for interest in unmanned maritime vehicles among Singaporean and Indonesian policymakers was the resurgence of small boat-based attacks, typified by the earlier attack on the USS Cole. Given the prevalence of both piracy and terrorist groups in Southeast Asia, this style of attack would be a significant concern to the security services of both Indonesia and Singapore. From a military standpoint, remote-operated unmanned maritime vehicles have been utilised primarily for surveillance and force protection. As of late 2019, there has not been major progress made by either Singapore or Indonesia towards designing a remote-operated or autonomous vehicle whose primary purpose is to be a surface combatant.[2]

Arguably the most influential aspect of UMV development is their potential to act as a 'force multiplier', improving the effectiveness and efficiency of state surveillance. This has been a driving factor behind their acquisition by militaries and state security agencies across Southeast Asia. Deployed surveillance USVs include the American Fleet Class Unmanned Surface Vehicle, which resembles a small motorboat and is used for short-range surveillance and early detection. As an example, consider the difference in manpower required to conduct surveillance patrols in a disputed littoral area between using multiple USVs under the supervision of a single human officer and, for example, a Thailand Navy M21-class patrol boat, which is operated by a nine-man crew (Parameswaran, 2018). Unmanned Underwater Systems are substantive more efficient for long-term surveillance, especially in the anti-submarine role. The Wave Glider, which is a two-part system that is designed for longer-term surveillance using acoustic sensors and passive sonar, is an example of a surveillance-oriented UUV.

Closely linked to their surveillance capacity, UMVs offer an increasingly effective alternative to surface vessels or aircraft for force protection. States have been forced to re-think their strategies for protecting vessels, particularly those that are making re-supply visits or deployments to ostensibly friendly foreign ports by the resurgence in the use of small, fast boats as suicide weapons, a tactic reminiscent of the fireships in the age of fighting sail. Because they can loiter in high-risk environments for hours at a lower cost under rotating operators, mobile

UMVs are seen as an effective solution to this threat. The Fleet Class USV is illustrative of the common characteristics of a force protection UMV, modelled on the chassis of a small speedboat with multiple sensors and limited autonomy (with a human operator supervising its patrol pattern). The B850 High-Speed Patrol USV is a Chinese platform designed to fulfil a similar role.

While based on their security environments and geographic factors, Southeast Asian states should emphasise procurement of maritime and aerial platforms; however, this would have challenged the dominance of army leaders within ASEAN militaries, which has previously limited spending and resource allocation to regional navies and air forces (Raymond, 2017). The proliferation to date of unmanned aircraft in the region and Indonesia's adoption of the Global Maritime Fulcrum strategy (which emphasised the need for greater naval capability modernisation), would indicate that policymakers have made progress in overcoming this traditional barrier.

4.3 Key actors in the post-demonstration point proliferation

Even after significant investment in recent years in military modernisation in the region, none of the ASEAN states has developed the capacity to domestically produce a UCV that could compete on the international export market with those produced by the United States, with the possible exception of Singapore. However, both Indonesia and Singapore have long-established track records of purchasing advanced military platforms where they lack the capacity to reasonably produce a comparative model. It is also interesting that the key arms exporters into Southeast Asia are also among the leading states in the development of unmanned combat vehicles, including the United States, China, Russia, South Korea and Israel. Within this region, it is becoming increasingly common for ASEAN member states to tie mandatory technology transfer provisions such as Indonesia's Defence Industry Law (2012) with agreements to procure advanced weapon platforms 'off the shelf', while other ASEAN members have entered into domestic development programmes with states outside of the region (such as Belarus) to improve their indigenous models.

The influence of arms-exporting states goes beyond merely selling platforms, especially when the platform is an emerging technology. In addition to the platform itself, procurement arrangements regularly include training, maintenance and access to spare parts, all of which enable the exporter to exert influence over the adopter's deployment of those systems. This section will demonstrate the role played by exporting states in the proliferation of UCVs and outline the limited efforts to date by great powers to influence their use by purchasing states in Southeast Asia.

4.3.1 USA

The United States is a major exporter of unarmed UCVs, which is reflective of their status as the leading developer of unmanned military systems and home to an advanced arms industry. However, despite rising demands from its allies and the occasional congress delegation, the United States has only authorised two sales of armed UAVs to before the end of 2018. In both cases, MQ-9 Reapers were transferred, first to the United Kingdom and second to Italy (Ewers et al., 2017), although Australia has reportedly purchased Reapers subsequent to this after a period of delay (Roggeveen, 2018). American allies that have become frustrated with the approval delays and export barriers for the US-designed armed UAVs, such as Jordan, have generally turned to China or Israel (Ewers et al., 2017). France's decision in September 2017 to arm its US-made, and initially unarmed, Reaper UAVs was potentially a response to this frustration.

Despite its leading role in their emergence, the United States has a comparatively less influential role in international efforts to emplace norms and regulations around the sale of armed unmanned platforms. The main regulatory framework that currently governs the exportation of unmanned aerial vehicles, armed and unarmed, is the *Missile Technology Control Regime* (MTCR). The MTCR is a legacy regulation, initially introduced in 1987, that relies upon voluntary compliance. Its original purpose was to restrict the proliferation of unmanned ballistic missile systems that could be used to deliver weapons of mass destruction. Because armed unmanned vehicles with a payload over 500 kg are considered Category I items, they are subject to an 'unconditional, strong presumption' of denying export authority, while many of the guidance systems and aeronautics components are considered Category II and require strong assurances that transferred components will not be on-sold. Aside from the United States, notable signatories include India, Russia and the Republic of Korea; however, none of the ASEAN member states are signatories. This protocol is generally considered to be one of the few internationally recognised methods for limiting the proliferation of UAVs, but it does not appear to apply to unmanned maritime vehicles or unmanned ground vehicles.

In a renewed effort to shape and limit the proliferation of unmanned systems, the United States has implemented two additional policies to which potential export partners must agree. These policies were effectively normative behaviour tools, attempting to establish a framework of international norms around the use of armed UAVs before they had fully diffused. The first was the 2015 *US Export Policy for Military Unmanned Aerial Systems*, which bound receiving states to only utilise US-produced UAVs within certain behaviour guidelines. The latest US-led effort was the 2016 *Joint Declaration for the Export and Subsequent Use of Armed or Strike-Enabled UAVs*, which has been signed by 53 states (Bureau of Public Affairs, 2016), including the Philippines and Singapore. This document encouraged signatories to abide by international law and conduct UAV operations with an appropriate level of transparency (Ewers et al., 2017). Adopting a normative framework to govern, regulate and shape unmanned aircraft exports is more likely to succeed than attempting to implement an outright ban of a technology that is rapidly proliferating through multiple state and non-state exporters.[3] However, the Joint Declaration has been criticised by academics and Amnesty International (2017) for setting standards too low, while other reports have questioned whether these efforts will be effective unless the United States is able to better capitalise on its existing arms-transfer arrangements to become the main exporter, gaining leverage to influence purchasing states (Ewers et al., 2017). This criticism is supported by the fact that Russia, China and Israel, none of which are signatories (Mehta, 2016), have proven very willing to sell armed unmanned combat vehicles to ASEAN member states.

4.3.2 *China*

The modernisation and rapid expansion of the Chinese military are often described in both the media and scholarly literature as the main reason for instability in Southeast Asia. While this downplays the impact of other regional security challenges, it does reflect the fact that regional middle powers are operating within an emerging hegemonic competition (Ikenberry, 2016). Within this competition, it is becoming increasingly clear that Chinese military technology is gaining on the United States in operational capacity and strategic reach (Allison, 2017). Chinese military development doctrine enshrines the idea that the Chinese and US military development is triggering a series of global Revolutions in Military Affairs and that, therefore, the People's Liberation Army (PLA) needs to accelerate its development efforts in 'domains of emerging military rivalry' (Raska, Undated), such as increasingly autonomous weapon systems and cyber warfare.

China has emerged as one of the two leading exporters of armed UAVs and, like their nearest competitor, Israel, are not signatories to the MTCR. In the comparative absence of the United States from the international armed UAV market, China has had considerable success marketing its Caihong-4 surveillance and strike UAV. The Caihong-4 has roughly comparable specifications to the MQ-9 Reaper and more than a passing resemblance. Per a June 2017 policy paper from the Centre for a New American Security, China has sold armed UAVs to Egypt, Jordan, Saudi Arabia, Iraq, Kazakhstan, Myanmar, Nigeria, Pakistan, Turkmenistan and the United Arab Emirates (Ewers et al., 2017). Another Chinese UAV that has been promoted for export is the Wing Loong II (also called the Pterodactyl), which is capable of being armed.

Chinese development of unarmed UAVs is also continuing at a rapid pace. Even in 2014, the PLA publicly displayed four new UAV models, and China subsequently stated an objective to acquire over 40,000 UAVs by 2023 (Office of the Secretary of Defense, 2015). Current Chinese surveillance UAVs range from tactical, short-range models (such as the ASN-206) to longer endurance models (like the BZK-005). China has also developed sophisticated High-Altitude Long Endurance (HALE) UAVs, such as the SYAC Divine Eagle. The most interesting development, however, has been the Sharp Sword (Lijian) armed combat UAV. The Sharp Sword is designed to deliver a first strike payload while protected by stealth features. It appears to be an answer to the British and American stealth unmanned combat aerial vehicle programmes, yet it is unclear at this point whether the Sharp Sword or its derivatives will be offered for export.

In addition to China's development and export of increasingly sophisticated armed UAVs, it has also made significant investment in developing unmanned maritime vehicles. A 2015 RAND Corporation report stated that the Chinese government had funded at least 15 research teams specifically to develop unmanned maritime platforms for military use (Chase et al., 2015). Furthermore, the People's Liberation Army Navy has access to remote-operated underwater vehicles with surveillance capabilities. In 2017, China deployed multiple unmanned underwater vehicle prototypes for 'scientific research' in the South China Sea (Chandran, 2017). Further committing to UUV development, the Zhuhai municipal government began construction in February 2018 on what it claimed was the world's largest unmanned maritime vehicle testing area (Long, 2018). Developing unmanned underwater vehicles would enable the People's Liberation Army Navy to project power or deny access without the same risk of escalation as committing manned vessels (Nurkin et al., 2018). A capability that would obviously be a serious concern for China's neighbours, particularly those ASEAN states affected by its claims to territory in the South China Sea. While there is little evidence that China has exported unmanned surface vehicles at the time of writing, this does not mean that exports will not occur in the future.

4.3.3 Israel

Israel has emerged as a major exporter of complete UAV systems and is Singapore's longest-standing arms export partner. Between 2005 and 2012, Israel was estimated to have exported $4.62 billion worth of UAV technology to 30 states, with the Asia Pacific being its second-largest market (IHLS Janes, 2013). For small scale, low endurance operations, Israeli UAVs include the IAI Heron, IAI Panther and Elbit Skylark I-LE. In terms of MALE UAVs, Israeli offerings include armed (IAI Eitan) and unarmed (Elbit Heron 900) aircraft. Israel also offers loitering attack munitions, or suicide drones, such as the Harpy. These explosive-tipped drones are designed to semi-autonomously track and identify enemy radar sites before dive bombing them.

Complementing their UAV exports, Israel is also a major developer of unmanned maritime vehicles. Israeli-made unmanned maritime vehicles include the Protector and the Seagull. The Protector was originally designed by Rafael Advanced Defense Systems and initially deployed by the Singaporean Navy in 2005. The Protector is remote-operated, armed and highly manoeuvrable. Its key purposes are surveillance and force protection. In 2017, a third-generation upgraded model demonstrated its ability to fire Spike ER missiles (Williams, 2017). Fulfilling a similar role with a greater emphasis on anti-submarine warfare, the Seagull (designed by Elbit) is capable of autonomous operation and able to deploy an autonomous unmanned underwater vehicle to aid its efforts to intercept and engage submarines.

Importantly, Israeli unmanned platforms are generally smaller and less complex than those built in the United States. There has been little public evidence that Israeli firms would be interested in committing to large-scale or high-cost platforms, such as stealth UAVs or unmanned surface combatant ships. The fact that Israeli defence companies are actively promoting unmanned maritime and aerial vehicles at trade shows indicate that they will continue to export to the Asia Pacific and, given that Israel is not bound by the MTCR and is the major arms supplier to multiple ASEAN states, it seems likely that the presence of Israeli-designed weapon systems in Southeast Asia is not likely to decrease.

4.3.4 Republic of Korea

Although the Republic of Korea (ROK) has not achieved the same export profile as the states described above, it has emerged over the last decade as a major arms supplier to Southeast Asian states. Given its ongoing efforts to develop increasingly autonomous unmanned ground, maritime and aerial weapon platforms, it would be remiss not to briefly engage with its role in remote-operated UCV proliferation. In addition to its advanced military technology, ROK is viewed more favourably by the ASEAN member states compared to its East Asian neighbours. This is because the ROK is considered to be more neutral than Japan, China or the United States, yet maintains influence with all three (Cronin and Lee, 2017). This perception increases the value of the ROK as a potential supplier of advanced weaponry for ASEAN states that are wary of offending one of the superpower states by purchasing too much from the other. It was also a key reason behind the ROK being invited to join the East Asia Summit.

The Republic of Korea has plainly benefited from the burgeoning arms race in Southeast Asia. Since 2009, South Korean arms exports have risen 1,100%, and the majority of purchases were in the Asia Pacific (Harris, 2016). Since 2007, South Korea signed 39 bilateral security agreements, and IHS Markit (owner of Jane's Defence) predicted that ROK defence export revenue would surpass China's by 2020 (Cronin, 2017). As an example, the major partner in the KF-X Future Fighter programme was Indonesia (Grevatt, 2017), although this particular deal has subsequently been under a cloud with the Indonesian Air Force reportedly considering Russian fighter jets. Existing South Korean UAVs include a tactical UAV developed by Korea Aerospace Industries and the Korea Aerospace Research Institute. While recently unveiled efforts in developing unmanned surface vehicles include the armed Haegeom USV, which is designed to patrol waters near the Northern Limit Line with limited autonomous navigation capability. As a signatory of the Missile Technology Control Regime, South Korea's ability to export advanced weapon systems is vulnerable to intervention by the United States. The 2015 intervention by the US to torpedo a potential deal to sell advanced T-50 trainer aircraft to Uzbekistan (Caverley, 2017) was illustrative of this potential. While there does not appear to have been any similar interventions in the UAV space, their MTCR obligations would be impacting which markets that the Republic of Korea defence industry engages with.

The United States, China, Israel and the ROK all major producers of advanced weapon systems with ongoing arms export relationships with ASEAN member states, through which they contributed to the development of unmanned combat vehicles and their rapid proliferation. The ability to purchase advanced unmanned platforms 'off the shelf' was a crucial enabler of proliferation into Southeast Asia, offsetting the lack of domestic capacity in the military-industrial base of key ASEAN states. Combined with developing operational praxes which have been emulated by secondary adopters, these actors have played a significant role in shaping how the mid-adopter ASEAN states have engaged with UCVs.

4.4 Applying adoption variables to UCVs in Southeast Asia

It is insufficient to argue that unmanned combat vehicles proliferated at such a rate in the 2010s purely because there was sufficient demand from smaller states to persuade advanced states to allow the export of complete, 'off the shelf', remote-operated weapon platforms. Intention must be paired with capacity, and therefore it is important to evaluate the capacity of early-majority and late-majority responding states to understand whether each state was able to acquire UCVs. The second half of this chapter will demonstrate how non-resource variables shaped the response of Indonesia and Singapore to the proliferation of unmanned combat vehicles to the extent that, while both states pursued a limited adoption strategy, this was largely limited to lower complexity platforms, which would still address their perceived capability requirements. This analysis offers an illustrative example of the effect that non-resource adoption variables could have on Indonesian and Singaporean reaction to the emergence of lethal autonomous weapon systems.

4.4.1 Security environment

The security posture of ASEAN members is reflective of the myriad traditional and non-traditional challenges that threaten regional security in Southeast Asia and has shaped how Indonesia and Singapore perceived the value of adopting UCVs. Major regional security issues include China's aggressive policy in the South China Sea, regional military modernisation and North Korean provocation. However, it is important to also consider the continued pressure non-traditional threats (such as human trafficking, piracy and terrorism) are placing on ASEAN states that have been operating comparatively ancient weapon platforms. Interestingly, Indonesia and Singapore have both indicated that they perceive a greater risk stemming from the threat posed by current non-traditional security issues to internal and regional stability than the more traditional risk of direct state aggression.

However, this is not to say that these militaries have abandoned their traditional concern with preventing territorial intrusion, deterring state aggression and projecting state power. The increasingly assertive, even aggressive, posturing by China in territorial disputes in the South China Sea has certainly played a role in promoting Southeast Asian military modernisation. As an example, consider the repeated clashes between Chinese fishing fleets and Indonesian authorities near the Natuna Islands, which have escalated to include standoffs between government vessels, the demolition of captured fishing vessels and prompted

Indonesia to establish a formal military presence and regional command in the area (Parameswaran, 2019).

Unfortunately for regional stability, the self-perpetuating aspect of concurrent military modernisation is particularly problematic in Southeast Asia due to the historical tension and political disputes between Southeast Asian states. For example, while Singapore has made improving defence cooperation and relations with Malaysia and Indonesia a priority since the early 2000s, the re-ignition of tensions around disputed territory near Tuas between Malaysia and Singapore in January 2019 (Rahmat, 2019) emphasised that fragility remains in the relationship.

In terms of non-traditional security issues, ASEAN member states (particularly Indonesia and Singapore) have been increasingly threatened by terrorism and organised criminal groups and pirates continue to operate in their territorial waters. Furthermore, ASEAN states, including Indonesia and the Philippines, have recently suffered from significant internal instability and even outright rebellion. These issues pose significant to member states' ongoing economic growth and political stability, which are vital to their broader military modernisation efforts.

Given that 2.5% of the world's ocean surface is encompassed within Southeast Asia, it would stand to reason that ASEAN states should prioritise the adoption of maritime and aerial unmanned platforms, although this has not always been the case (Raymond, 2017). Unmanned platforms are notably more resource-efficient for active surveillance and can be deployed without human risk into dangerous or difficult-to-navigate sections of coastal waters to interdict territorial intrusions or to track and intercept the movement of militants, pirates and contraband. Remote-operated platforms allow multiple operators to rotate through piloting a single UAV, improving its capacity for long-term surveillance while somewhat offsetting the impact of boredom, distraction and fatigue. Finally, remote-operated platforms can be used to assist human law enforcement or military personnel to respond safely to ongoing terror incidents, reducing overall casualties in situations that are too dangerous for first responders, the obvious example being the long-standing use of remote UGVs for explosive device disposal.

The need for a regional response to transnational security issues and inspired by their use by western forces and China, ASEAN member states have begun efforts to acquire and utilise UAVs. So far, the maritime focus of their security environment has contributed to six ASEAN nations developing the capability to produce or acquire small-medium size surveillance UAVs (Indonesia, Singapore, Thailand, Malaysia, Vietnam and the Philippines).

4.4.2 Resource capacity

In response to these issues, there has been a marked increase in military spending across the region (Heiduk, 2017), funded by dynamic economic growth across the region. Southeast Asia is one of the fastest-growing regions in terms of economic growth, ahead of Latin America and Africa, with a regional average of 5% growth over the last five years (OECD, 2018). Driven by regional distrust and the increasingly belligerent territorial claims by China, the overall trend among ASEAN states has been to increase their defence spending at a regional average rate of 9% since 2009 (Laksmana, 2018). Singapore maintains the largest military expenditure in Southeast Asia, with a total defence budget of SGD 14.2 billion (US$10.2 billion) in 2017 (International Institute for Strategic Studies, 2018), while Indonesia's defence budget is the second largest, reaching 120 trillion Indonesian Rupiah in the same year (US$8.98 billion) (International Institute for Strategic Studies, 2018). The significance of Singaporean and Indonesian defence spending is further illustrated by the fact that 50% of defence imports in the region are destined for these two states (Dowdy et al., 2014).

While there are some states that have not followed suit, the overall trend among ASEAN states has been sustained increases in defence spending, a trend that has not gone unnoticed by the international military industry. In 2016, 43% of global arms imports were destined for the Asia-Oceanic region (Fleurant et al., 2017) and Southeast Asia was collectively the second-largest military import market between 2007 and 2012. Since 2009, ASEAN spending on defence imports has spiked by 71%. Between 2000 and 2010, arms exports to Malaysia rose by 722% and those to Indonesia by 84% (Simon, 2012). Vietnam's arms purchases rose 699% between 2011 and 2015 (Parameswaran, 2016).

These increases in defence spending have been focused on modernising military equipment, especially upgrading or replacing ageing major combat platforms. For example, Southeast Asian air fleets were characterised by aircraft acquired in the 1970s and 1980s, and have been targeted for modernisation. Singaporean and Indonesian combat aircraft averaged at 16 years old until modernisation efforts saw the former invest $2.43 billion in modernising its F-16 fleet and the latter joined the ROK's KF-X Future Fighter programme as a major investment partner.

The development of domestic capacity to produce remote-piloted unmanned platforms among ASEAN member states (bar Brunei) has largely, but not exclusively, occurred through military technology transfer and dual-investment agreements with traditional trading partners. Levels of success certainly vary, while Vietnam was successful in its efforts to develop a competitive indigenous, high-endurance UAV with the support of Belarus (Thayer, 2018), Indonesia's domestic arms producers are still in the process of converting recent increases in resource allocation into greater capacity to produce internationally competitive UCVs. Despite entering production in 2004, the indigenously produced Indonesian Wulung UAV only entered production in 2013 and possessed limited endurance and payload capacity (Parmar, 2015). The Indonesian Air Force allocated more than US$16 million in 2011 to procure for UAVs from the domestic company PT Dirganatra (Taylor, 2011). Singapore has enjoyed greater success, with the ST Aerospace-designed Skyblade family of tactical level UAVs being issued to army units (Desker and Bitzinger, 2016), and the 3D-printed UAV developed by civilian commercial company O'Qualia (Parmar, 2015).

Indonesia, Singapore, Myanmar and Thailand have also purchased remote-piloted aircraft and related technologies 'off the shelf' through existing arms agreements with states that are known to be developing increasingly autonomous weapon systems. Examples include Indonesia's purchase of Heron II UAVs from an Israeli firm, which were intended to be the first of 80 foreign-purchased UAVs by 2017 (Rahakundini, 2016). The Singapore Armed Forces (SAF) also purchased Heron UAVs as well as Protector USVs, which were subsequently deployed by their Navy. This appears to be part of a broader pattern of ASEAN states capitalising on existing arms agreements to procure unmanned, remote-operated, platforms from foreign powers to cover capability gaps until their domestic production capacity advances.

4.4.3 Organisational capital capacity

The second adoption variable for consideration is whether the state possessed sufficient organisational capital capacity to incorporate unmanned combat vehicles into their power projection apparatus. Horowitz (2010) describes three tests for measuring a state's organisational capital capacity; critical task focus, level of investment in experimentation and organisational age. In the case of remote-operated unmanned combat vehicles, Indonesia and Singapore diverge in these variables, but not to the level seen in the case of autonomous weapon systems. Despite this divergence, both states demonstrated a preference for aerial and maritime unmanned vehicles that could be deployed in border security and surveillance roles.

4.4.3.A Critical task focus

The critical task focus of the Indonesian and Singaporean militaries shifted procurement efforts in the mid-2000s towards acquiring and improving platforms that responded to internal, non-traditional and non-state threats. The Singaporean Ministry of Home Affairs explored the use of unmanned platforms for law enforcement purposes, while the Navy's adoption of the Protector USV reflected a concern for the safety of military and commercial vessels in harbour. From the Indonesian perspective, the 2015 Defence White Paper reaffirmed the military's commitment to responding to non-state threats and, in the same year, the Ministry of Transport issued a regulation that explicitly allowed for the use of UAVs for border and maritime patrols (Rahakundini and Prasetia, 2016). The Indonesian navy has also financed the development of 'kamikaze' UAVs to be deployed against illegal fishing vessels (Cassingham, 2016). This internal focus is not unusual among Southeast Asian states, whose naval vessels spend the majority of their deployed time supporting internal security agencies to police territorial waters and contributing to multilateral efforts to combat regional non-traditional security threats (Laksmana, 2018).

4.4.3.B Level of investment in experimentation

Unsurprisingly given its powerful, advanced economy, strong commercial research and development sector and long-standing commitment to maintaining a 'secret technological edge' over its neighbours (Matthews and Yan, 2007), the Singaporean Armed Forces further outpaces their Indonesian counterparts in the level of resources consistently committed to experimentation. Singapore's defence technological community includes the major civilian developers such as the conglomerate Singapore Technologies Engineering (STE) and its component companies, the Defence Science Organisation National Laboratories (which focuses on defence research and development), the Defence Science and Technology Agency (which coordinates the SAF's innovation, development and procurement processes), and the Future Systems and Technology Directorate (which develops innovative operational concepts). The latter is the product of a 2013 merger of the Defense Research and Technology Office and the Future Systems Directorate, which was given 1% of the total defence budget in its first year to challenge the SAF's exiting strategic thinking. These agencies each played a significant role in the adoption of remote-operated platforms as part of the Third Generation SAF (Bitzinger, 2018).

Indonesian interest in UAVs started with the commercially designed Sky-Spy-5 in 2003 (Rahakundini and Prasetia, 2016), seven years before the establishment of the Defence Industry Policy Committee, or *Komite Kebijakan Industri Pertahanan* (KKIP), which would become the guiding agency for its military modernisation efforts. The main Indonesian military research body is the Ministry of Defence's Research and Development Agency (*Badan Penelitian dan Pengembangan*, also known as *Balitbang*). Government research is supported by a slowly developing domestic military industry, which has demonstrated a capacity to successfully integrate military technology transfers from foreign weapon platforms. While Indonesia's domestically produced drones are technologically inferior to those sourced from Israel, there are multiple civilian and state actors actively participating in further development.

4.4.3.C Organisational age

Theoretically, the more advanced the organisational age of a given military, the more resistance would be encountered in an attempt to adopt an innovation. Horowitz (2010) proposes two measures, the first being the length of time since the state lost a major war or underwent regime change, the second is the nature of civil–military relations in a given state.

This first variable is not very applicable in the case of Southeast Asia because, despite the ongoing regional tensions and inter-state territorial disputes, the ASEAN member states have quite remarkably avoided major conflict over the past 20 years, although some (such as Singapore) participated in the western-led Global War on Terror and others have suffered from internal conflicts (including Indonesia, Myanmar and the Philippines). Similarly, Singapore's People's Action Party (PAP) has consistently retained power since independence, although the Singaporean military has undergone three major evolutionary changes. On the other hand, while Indonesia renewed its interest in UAVs and establishing a green water capability following the election of President Widodo; overall the post-Suharto de-politicisation of its military has not been completely successful, and the TNI remains a powerful force in domestic politics.

There are substantial differences in the civil–military relationship between Indonesia and Singapore. The Indonesian military remained top-heavy and committed to its Total People's Defence strategic doctrine in the face of the civilian-led Minimum Essential Force and Global Maritime Fulcrum strategic concepts, which would have placed greater emphasis on naval and air assets than is typical in the army-dominated TNI. As a result, the majority of the UAVs operated by the TNI are short- and medium-endurance variants designed for surveillance, with seemingly little interest in strike capability. As noted by Rahakundini and Prasetia, further engagement with UCVs is blocked by a lack of funds (which are largely spent on personnel costs for the army) and a lack of political will (Rahakundini and Prasetia, 2016). Contrastingly, Singapore's civil–military relationship is heavily slanted in favour of civilian leadership (Raska, 2015), who have embraced the use of remote-operated vehicles as a cost-effective solution for their small population and a useful way to promote further, economically lucrative, commercial development of related technologies (Desker and Bitzinger, 2016). Singapore's more advanced organisational capital capacity resulted in a greater willingness to experiment with and acquire UCVs compared to Indonesia.

4.4.4 Receptiveness of domestic audience

Unfortunately, there are no published statistics illustrating public opinion towards increasingly autonomous or remote-operated unmanned combat vehicles in Indonesia. None of the major surveys conducted as of the end of 2018 had a significant number of respondents from ASEAN member states. However, there are other sources of data that can inform this variable. For example, a 2014 survey by the Pew Research Center found that 74% of Indonesians opposed the United States' use of remote-piloted UAVs for targeted strikes (Pew Research Center, 2014). However, this is balanced by evidence of support from the Indonesian government and defence contractors. For example, the Indonesian Ministry of Transport released a regulation in 2015 that explicitly allowed the state to utilise UAVs in roles ranging from border patrols to weather observation (Rahakundini and Prasetia, 2016), while PT Dirgantara Indonesia (PTDI) announced that it would be collaborating with Turkish Aerospace Industries to develop UAV platforms as recently as January 2018 (Desk, 2018).

Public opinion data is also lacking in the case of Singapore; however, unmanned platforms have been a major component of the third-generation SAF with support noted in scholarly, commercial and government research papers. Furthermore, the Singaporean Ministry of Transport is the head of a multi-agency task force whose purpose is to promote the 'innovative use of unmanned aircraft' in both the private and public sectors (Desker and Bitzinger, 2016). The FSTD has also been credited with playing a major role in developing the Airspace Management Technology (Raska, 2015), which is a necessary step for the safe commercial use of UAVs in Singapore's crowded airspace. In 2015, Singapore's Parliament discussed the risks and benefits of further promoting the commercial use of UAVs through the *Unmanned Aircraft (Public Safety and Security) Bill 2015* (Desker and Bitzinger, 2016). Finally, it is unlikely that Singapore would risk the substantial economic benefits of robotics, artificial intelligence and unmanned systems with overly restrictive regulation of civilian commercial developers (Desker and Bitzinger, 2016).

Despite the lack of published data on public opinion towards remote-operated military platforms in Indonesia and Singapore, there is evidence from both states of notable commitment to remote-operated systems by military, political and commercial actors. This supported the limited adoption of unmanned aerial vehicles by state agencies beyond the military, and both states committed significant resources to improve the capacity of their domestic industrial bases to participate in the rapidly expanding civilian market for unmanned aerial vehicles.

4.4.5 Adopt specialised operational praxis

The development or emulation of a specialised operational praxis is essential for a military to successfully integrate a given innovation. In the case of unmanned combat vehicles, Southeast Asian states have followed their developer-state peers in exhibiting a preference for deploying remote-operated systems in the aerial and maritime domains over ground-based platforms. Singapore is one of the few states in the region which has expressed an interest in UGVs beyond the prolific ordinance-disposal robots, flagging an interest in utilising UGVs for battlefield logistics and casualty recovery. This is unsurprising given the SAF's documented interest in learning from the US doctrine (Laksmana, 2017).

In the case of unmanned aerial vehicles, there has been a clear recognition among most ASEAN member states of the benefits UCVs offer for surveillance at both the tactical and strategic levels. While it is commonly acknowledged that Southeast Asian states would lack the informational infrastructure to sustain an intercontinental deployment of unmanned aircraft in either a strike or surveillance role (Ewers et al., 2017), there has been widespread adoption of UAVs for border surveillance and limited integration into counterpiracy efforts. As an example, Indonesia's interest in developing UAVs has favoured low- and medium-endurance models, which are less technically demanding while still possessing the capabilities required for its internally oriented Critical Task Focus. Interestingly, however, there was little apparent interest among ASEAN member states in arming their developing UAV stocks. The major exception to this, Myanmar's decision to purchase the strike-capable CH-4 Caihong-4 UCAV from China (Ewers et al., 2017), demonstrates that the lack of armed UCV adoption by the other ASEAN member states is not necessarily the result of a lack of resources. This conclusion is further reinforced by the instances of non-state actors in Syria, Iraq and Ukraine, utilising modified civilian remote-piloted aircraft for kinetic strikes. In the case of Singapore, the decision not to outwardly pursue armed unmanned platforms reflects a long-standing aversion to adopting weapon systems that would be seen by neighbouring states as aggressive (Scharre, 2018).

Given the centrality of maritime territorial disputes to the security environments of Singapore and Indonesia, the adoption of an operational praxis for the deployment of unmanned maritime vehicles has emerged as an understandable priority. For example, Singapore adopted the Israeli-made Protector USVs, which can be used for both surveillance and force protection, and subsequently deployed them in a limited counterpiracy role (Desker and Bitzinger, 2016). Furthermore, there are clear operational benefits to be achieved from deploying remotely operated maritime vehicles for surveillance in waters that are too dangerous or difficult to navigate for the reliable deployment of manned vessels. Finally, the potential to utilise remotely operated platforms for area denial and force protection are both appealing to ASEAN militaries who have identified that the threat of violent non-state actors as their main security priority.

4.5 Outlining ASEAN member state engagement with UCVs

Overall, analysing the proliferation of unmanned combat vehicles in Southeast Asia through the prism of the above diffusion variables demonstrates that Indonesia and Singapore were able to achieve a sufficient resource capacity to pursue a strategy of limited engagement with UCVs, while their security environments and organisational capital capacity informed a preference for unarmed aerial and maritime platforms.

Although it first experimented with unmanned technology in 2003, Indonesia's adoption of unmanned combat vehicles is still at its early stages. The key barrier to further development or acquisition of unmanned systems has been a lack of clear strategic direction, a conservative senior military leadership and the long-standing allocation of a significantly lower percentage of GDP to defence than the regional average. However, the benefits, resource efficiencies and the prestige associated with acquiring unmanned platforms all indicate that Indonesia is likely to pursue unmanned combat vehicle technology as part of its broader modernisation goals.

For Singapore, with its restricted geographic footprint and ageing population, maintaining a military technology offset is considered the great force multiplier against rival states. The adoption and development of unmanned weapon systems was a core component of Singapore's third-generation transformation. The adoption of remote-operated platforms also influenced the shift from the 'porcupine' doctrine, which essentially advocates deterrence based on perceived ability to outlast an invader through attrition and forward defence (Desker and Bitzinger, 2016), to the smart-power based 'dolphin' strategy (Tan, 2015). Unmanned platforms allow Singapore to offset its smaller population, increase the combat effectiveness of its comparatively small military and enable lower-cost forward defence in the event of inter-state conflict. Interestingly, Singapore has not demonstrated a strong interest in acquiring armed UAVs, however, SAF would have the capacity to rapidly adapt current platforms in the event of a significant security incident or threat to its territory.

Beyond Indonesia and Singapore, the overall response of ASEAN member states has been variations of a similar scale limited adoption. This adoption varies in scope, source and extent between member states; for example Malaysia produces four variants of short- and medium-range UAVs, while Thailand purchased Elbit Hermes 450 medium UAVs from Israel, Vietnam partnered with Belarus (Gady, 2015) to develop a respectable domestic UAV production capability (Thayer, 2018) and Myanmar purchased CH-4 Caihong-4 surveillance and strike UAVs from China (Ewers et al., 2017). Adoption has not been a uniform response; for example Brunei has not adopted military UAVs, although the Crown Prince was given a tour of the Republic of Singapore Air Force's Unmanned Aerial Vehicle Command during a state visit in October 2018.

The analysis in this section partially supports the prevailing contention in exist-ing scholarly literature that ASEAN member states do not possess the resource capability to procure or produce comparably advanced platforms or maintain the sophisticated C4ISR and data infrastructure (for transfer and storage) necessary to adopt a comparable operational praxis to that of the United States or China. However, examining the critical task focus and security environment of these states demonstrates that neither Indonesia nor Singapore would have been well served by platforms with this level of capability. Nevertheless, this should not be used as an excuse to dismiss the potential impact of even tactical level use of remote-operated weapon platforms in the South China Sea.

The widespread adoption of unmanned aerial vehicles in Southeast Asia is demonstrative of the disruptive element of UCVs, which is shared by increasingly autonomous weapon systems. That as an inherently dual-use innovation diffuses, smaller states and violent non-state actors that do not require the high-level capa-bilities of advanced US military grade systems have a variety of alternatives for acquiring a platform of lower, but sufficient, capability through domestic produc-tion, foreign partnership or simply buying from the civilian market.

4.6 Conclusion

In the modern geostrategic climate, the Southeast Asian region hosts one of the most important and concerning flashpoints for inter and intra-state conflict. Yet these risks are combined with immense economic potential. The result is a region where states are developing towards middle power status, with the economic and political growth that entails, under the shadow of ongoing hegemonic tensions between the existing superpower and a rapidly strengthening rival. This conflu-ence of events has sparked justifiable concern among security academics, military personnel and policymakers.

This chapter has argued that the rapid proliferation of remote-operated unmanned combat vehicles, the precursor innovation to AWS, offers a cru-cial insight into how regional state actors, including Indonesia and Singapore, are likely to respond to the subsequent emergence of increasingly autonomous weapon systems. Thus far, ASEAN states have enjoyed low-adoption barrier access to remote-operated unmanned vehicles and have modelled their organisa-tional integration on the approach taken by larger states outside the region. While the majority of the regional military modernisation spending has been invested in upgrading existing military assets and production capacity, existing arms-transfer relationships have offered some ASEAN member states ready access to remote-operated platforms without significant initial investment. This has been complemented by a rapidly expanding civilian commercial market that has already been co-opted by state and non-state actors, which has allowed states with lower resource capabilities to instead purchase COTS platforms.

As with previous disruptive technologies, as the underlying technology matures the unit cost will fall and diffuse and it is into this environment that unmanned military platforms are proliferating, a path that will also be followed by the first generation of lethal autonomous weapon systems. Therefore, the broader purpose of this chapter was to demonstrate how state actors responded to the emergence of unmanned combat vehicles as a military innovation, itself the precursor innovation to autonomous weapon systems, with a particular focus on Indonesia and Singapore as leading actors in Southeast Asia. The analysis within this chapter feeds into the book's broader enquiry into how the emergence of lethal autonomous weapon systems, a paradigm-shifting military innovation, will impact the security and stability of Southeast Asia.

Notes

1 These states were Egypt, Iraq, Jordan, Kazakhstan, Myanmar, Nigeria, Pakistan, Saudi Arabia, Turkmenistan and United Arab Emirates (Ewers et al., 2017).
2 Ongoing research efforts outside Southeast Asia include the Ocean 2020 programme (funded by the EU) and China's D3000 prototype.
3 Horowitz, M. quoted in Mehta (2016).

5   The rise of lethal autonomous weapon systems

5.1 Introduction

While the development of lethal autonomous weapon systems (LAWS) has relatively recently become the focus of public interest and scholarly concern, developing technology capable of operating with limited or no human involvement reflects a longstanding idea in both science fiction and real-world policy. Ignoring for a moment the significance of their impact, a Revolution in Military Affairs (RMA) still shares definitional elements with other forms of disruptive military innovation. This chapter argues that, while a LAWS demonstration point is not currently imminent, the development of its 'hardware' and 'software' components is already underway. This chapter offers important insight into how major states are acting during the current incubation period, as well as how their participation would affect a future demonstration point.

This chapter opens with a succinct exploration of each of the three classifications of autonomous weapon systems, as well as how AWS overlap with unmanned platforms and the military application of artificial intelligence.

This is followed by the main analytical component of this chapter, which contends that both components of LAWS remain underdeveloped at this stage. Beginning with the hardware component, this chapter will demonstrate that current technology would not allow for the safe deployment of a platform with complete autonomous control over its target identification and selection, but that a basic level of operability can be achieved in the other critical functions of movement through the battlespace and target engagement. This will be followed by a section that outlines some of the most prominent operational concepts that are visibly being developed to identify common themes and approaches.

The third section of this chapter offers a short comparative analysis of the progression made by both hegemonic competitor states in the region towards developing Lethal Autonomous Weapon Systems. In addition to comparative analysis, this section will outline how each state is responding to two major barriers to adopting LAWS as a first mover: developing and securing top-level expertise and maintaining adequate access to the relevant data sets required to train AI-enabled systems. Although this book is focused on regional middle powers, it is important to understand how the United States and China are approaching increasingly

DOI: 10.4324/9781003172987-5

autonomous systems. It is important to note that neither state currently admits to developing autonomous weapon systems and, while both have funded related research activities, currently maintain policies that would retain a human on the loop. Beyond the direct impact on their hegemonic competition, the investments by both states in AI and AWS are already influencing the perceptions of key ASEAN member states.

This chapter closes by outlining the involvement of four additional extra-regional states that are active within Southeast Asia and are publicly developing AI and increasingly autonomous systems for military purposes. The unique characteristic of AWS is the extent to which adoption would not necessarily require exotic materials, advanced manufacturing apparatuses or specialist knowledge. At its most simplistic, an AWS is a computer that analyses data input from multiple conventional sensors to inform its actions. This chapter will demonstrate how the comparative lack of these traditional acquisition chokepoints has already prompted greater participation by both state and non-state actors in the development of autonomous military technology. While a demonstration point would not be imminent, it is important to understand the role each of these states are playing in the current incubation period, both in anticipation of a future demonstration point and as an acknowledgement that AMT development is already influencing state behaviour in Southeast Asia.

5.2 'Autonomous' weapon systems, unmanned platforms and artificial intelligence

Given the ongoing debate over the definition of autonomous weapon systems and the fact that the disruptive element of this innovation is a capability rather than a discrete weapon platform, it is important to start with an outline of the pertinent definitions and distinctions in this space. While the majority of systems referred to in this book could also be characterised as 'unmanned', the distinguishing characteristic of autonomous weapon systems is that they can exercise a level of independent control over their 'critical functions'. Narrowly focused AI[1] is arguably the most important underlying technology for this innovation, enabling a LAWS to independently act within the battlespace, based on its interpretation of sensor data taken from its surroundings.

At the time of writing there have been no publicly acknowledged deployments of fully autonomous weapon systems. This is largely due to the ongoing legal and definitional uncertainty, although a genuine question remains as to the feasibility of imbuing a weapon system with capabilities that could be objectively classed as 'autonomous' (Anderson, 2016).While there have been deployments of weapon systems that have the capacity to operate in a manner independent from human supervision, the DoDaam Super Aegis II is an example (Parkins, 2015), a division must be drawn between whether these weapon systems are truly 'autonomous' weapon or merely 'highly automated' (Anderson, 2016). Despite the continued definitional inconsistencies and debate (detailed in Chapter 2), it is possible to broadly distinguish semi-autonomous and supervised unmanned platforms from a 'full' LAWS based on a function-based, platform-focused approach.[2]

Also known as 'human in the loop' platforms, semi-autonomous weapon systems are human-activated with a limited capacity to autonomously manoeuvre and/or engage designated categories of target within geographic limitations (ICRC, 2014). Despite superficial similarities, semi-autonomous weapons are functionally different from automatic weapons (like landmines), which merely react to a particular stimulus, and remotely operated unmanned platforms. Rather, a semi-autonomous weapon system is capable of independently distinguishing between potential targets and operates without direct human control within its pre-defined boundaries.

The Mobile Detection Assessment and Response System – External (MDARS-E) was a prime early example of a system in this category. The MDARS-E was developed by the Space and Naval Warfare Systems Centre (SSC Pacific), San Diego. The MDARS-E was able to autonomously patrol within an assigned territory (such as a fenced facility) (Mullens et al., 2003). Upon detecting an intruder, it gave an audible warning to turn back. If the intruder did not leave the guarded area, then the MDARS-E had the capability to engage with its pepper-ball (paintballs containing tear gas) gun (Carroll et al., 2002), while calling for assistance from human guards. It also had the ability to independently launch an onboard UAV to pursue a fleeing intruder. In 2005, the MDARS-E successfully completed a 12-month Operations Assessment and Early User Appraisal at Hawthorne Army Depot Nevada, the largest army munitions depot in the world (Shoop et al., 2006). The MDARS-E was subsequently deployed to guard US nuclear facilities in 2010 and an upgraded version is in development.

Taking a step further, supervised (human on the loop) platforms are capable of selecting and attacking targets independent of human command yet include a mechanism that allows a human supervisor to interrupt or terminate the weapon's engagement process within a limited timeframe. Unmanned platforms with these limited levels of autonomous control are the main category of autonomous military technology that are publicly under development. Supervised autonomous weapon systems are most commonly deployed in defensive roles, such as Close-In Weapon Systems (CIWS), which passively scan for incoming threats to the host vessel when in automatic mode. Upon detection of an incoming threat the human supervisor is alerted and, unless overridden, the weapon system engages the threat. The Russian Uran-9 unmanned ground combat vehicle (UGCV), which was deployed to Syria in 2018, and the Fleet class USV are both further examples of armed supervised weapon systems.

Finally, it is important to note that direct military applications of AI and other related technologies comprise only a comparatively minor section of the broader research efforts in these fields. In a reverse of the traditional development burden of an emerging major military innovation, development is primarily occurring outside of the security space. Rather commercial and university-based research has been principally intended to contribute to civilian projects, such as self-driving cars and home automation. As dual-use technologies, advances in related enabling components are still relevant in outlining our progress towards a future demonstration point of LAWS. However, in addition to the fact that AI software requires task-specific data, military co-option of these technologies would require far more robustness and resistance to interference than is generally present in civilian-designed systems.

5.3 Development towards a LAWS demonstration point

The purpose of this section is to assess progress towards a demonstration point, where a first mover reaches the requisite level of autonomy and operational integration to demonstrate the capability to deploy an unmanned platform that reaches the working definition used in this book. Although there is no denying that a first mover will maintain their advantage over a fast follower, the status does offer significant opportunities, not least of which is the opportunity to convert this early development lead into an enduring influence over the future use of that military innovation.

While the 'hardware' component of a major military innovation typically attracts more public attention, an innovation is not simply a new weapon system or formulation for armoured plate, rather the technological breakthrough must be paired with organisational change. A basic analogy would be that a desktop personal computer makes a decent coffee table, but it requires an operating system for the user to fully access its potential. This section will demonstrate that, despite the efforts of various developmental actors, it is not yet technologically feasible to reliably deploy a LAWS that would meet the definition used by this book. Therefore, it remains an 'incomplete' innovation that does not yet support a demonstration point. This is not unusual in a major military innovation, for the reason that the time between the development of one component and the other is not set; it could occur immediately alongside the invention (as with nuclear weapons), within a few years of the technology maturing (such as with the first UCAVs) or extend decades (in the case of armoured warfare). Even though advanced lethal autonomous weapon systems may remain out of reach for smaller states and non-state actors in the foreseeable future, it is equally clear that LAWS-developing states are focusing on weapon systems that maintain some level of human input into the Observe, Orient, Decide, Act loop in the near-term. It is worth noting, however, that the maintenance of human input in a system would disqualify it as a fully autonomous system under the definition used in this book.

5.3.1 'Hardware' – Progress towards autonomous control over critical battlefield functions

The main difficulty in evaluating technological progress towards deployable fully autonomous weapon systems is that 'autonomy' is a non-binary capability (Anderson, 2016), it is not a stand-alone or easily identifiable weapon platform. Indeed, Horowitz recently described AI, arguably the most important enabling technology for LAWS, as closer to the steam engine or electricity than prior self-contained major military innovations (Horowitz, 2018). The key is in the detail of which functions that an unmanned system can operate autonomously; this section presents three approaches for evaluating this capacity. The first focuses on the platform's ability to *sense, decide* and *act* independent of a human operator. The second focuses on self-mobility, self-direction and self-determination. Finally, the third approach measures the platform's level of independent control over its 'critical functions', which are the processes used to 'select (i.e. search for or detect, identify, track, select) and attack (i.e. use force against, neutralize, damage or destroy) targets without human intervention' (ICRC, 2014). Based on these approaches, a fully autonomous weapon system would need the capacity to maintain a high level of independent control over its key functions: movement, target identification and target engagement. Therefore, this section evaluates the extent to which it is technologically possible for a platform to operate with autonomous control over these critical functional areas.

5.3.1.A Movement

It is unsurprising that this functional area is the most mature given that a significant requirement for an unmanned platform to autonomously manoeuvre through the battlespace (across all three domains) is the capacity to interpret data from sensors that are largely identical to those used in civilian robotics. Three of the most commonly used sensor types in unmanned platforms are GPS (or other forms of satellite positional navigation) (Ryan, 2018), LIDAR (Mukhtar et al., 2015) and computer vision technology (Development, Concepts and Doctrine Centre, 2018); which are also the primary sensors used for autonomous navigation in civilian systems. Unmanned platforms process data from these sensors to build an understanding of the environment around them, a mathematical map, which is updated as the device manoeuvres. While the clear majority of research in related fields, such as robotics, computer vision and human–machine interaction, is focused on civilian innovation (e.g. self-driving cars and indoor flight capability for civilian quadcopters), there is little difference in the sensors or the interpreting software.

'Follow me' platforms utilise the most straightforward variety of automated manoeuvre capability. These platforms typically rely on computer vision and lidar to maintain a connection with an assigned 'leader' and to independently avoid obstacles (Ivanova et al., 2016). However, this is a very low-level autonomous capability that is generally not able to identify a new leader (if the original was disabled). Despite its limited potential, 'follow me' movement capability is a convenient and effective way to bring heavy firepower or additional supplies to a small unit of soldiers, without the need to allocate limited mental resources to actively controlling a remote platform.

Aside from collaborative-reactive swarming, checkpoint-based navigation is the most common method for independent manoeuvre in the aerial and naval domains. This is based around pre-designated (or remotely assigned) waypoints (Sayler, 2015). The unmanned platform draws on data from onboard sensors and collision-avoidance software to autonomously manoeuvre through the battlespace, independently altering its route to increase efficiency, respond to changing objectives, or to avoid obstacles and potential threats (Kearn, 2018). A powerful indicator of the sophistication of waypoint navigation for aerial and maritime platforms is that effective versions are common in affordable civilian hobby UAVs. However, this is significantly more difficult to achieve for ground vehicles because their environment is more actor dense and subject to rapid change.

5.3.1.B Target identification and selection

The intense scholarly scrutiny that has been placed on the prospect of enabling an unmanned weapon platform to autonomously engage a target, the decision to pull the trigger, reflects a misunderstanding of the issues. The core distinction of 'autonomous' weapons lies in their capacity to 'undertake' the process of identification, rather than merely to respond to a particular stimulus (Anderson, 2016). Without the capacity to reliably identify and select legitimate (and, arguably more importantly, illegitimate) targets there is a danger that human supervisors would operate on the basis of overly enthusiastic interpretations of the platform's capability, even where 'meaningful human control' is theoretically maintained (which itself presents an inconsistency with existing IHL) (Anderson, 2016).

The effectiveness of current technology for reliably identifying targets varies dramatically between vehicular, structural and human targets. Vehicular and structural targets are easier for autonomous systems to recognise with visual cameras and image recognition software. While current technology can identify that a given object is human, this is largely based on their shape. The platform would be unable to reliably distinguish individuals in real time without human involvement, especially in a complex or unconventional battlespace. Despite its prominence in media accounts, using real-time facial recognition has serious reliability problems outside of sterile laboratory conditions (Development, Concepts and Doctrine Centre, 2018). While computers can identify basic behaviours (such as walking), they cannot intuitively leap from observing a behaviour (walking, running, putting hands up) to an inferred intention (surrendering) or a deduced conclusion (setting up an ambush) (Sparrow, 2015). This is particularly limiting considering the risk that combatants may just learn to avoid the behaviours that identify them as legitimate targets, or simply pretend to surrender, until the platform has moved on to another area of the battlespace. Furthermore, even if we disregard reliability issues and the black box problem, current machine learning techniques mean that training AI requires significant amounts of relevant data, a process that would require developers be given access to classified databases of active targets.

Overall it is clear that technology has not developed to a point that it would be feasible to deploy a weapon system with full autonomy over its target identification and selection process into a ground combat role without accepting a high level of risk to non-combatants and friendly personnel (ICRC, 2016). This position is supported by the fact that the vast majority of UGCVs and sentry guns currently retain human oversight (Boulanin and Verbruggen, 2017). However, given the limited capabilities of sensor and processing technologies, these human supervisors must exercise meaningful human control over their weapon systems, rather than merely relying on the target identification by unmanned platforms at face value.

5.3.1.C Engagement

The third function, selection of a method of engagement and target persecution, is well within the capacity of modern technology. Recall that a LAWS is an advanced unmanned *platform* with autonomous control over its critical functions, not a completely stand-alone weapon system (Horowitz, 2016). Indeed, allowing unmanned platforms to autonomously engage targets designated by human operators could offer significant advantages in terms of accuracy, reliability and reaction time over human-directed engagements. This is the logic underpinning the Loyal Wingman programme, which would pair a supervised wingman with a human lead-pilot, with the latter retaining responsibility for designating targets.

Consider the following scenario. A modified (so as to have autonomous target engagement capability) MUTT (an armed UGV) is operating with a squad of United States Marines in an urban counter-insurgency environment, when the squad comes under accurate sniper fire from multiple windows of a nearby apartment complex. The supervising marine utilises the MUTT's onboard cameras to designate the shooter as a legitimate target from the safety of cover. Unlike human soldiers, the MUTT is not affected by adrenaline or the need to avoid being shot and has the benefit of audio direction finding and precision tactical radar,[3] which allows it to rapidly and accurately track the trajectory of incoming shots as the shooter moves between windows. It quickly becomes clear that, in this scenario, a human-directed LAWS could effectively engage the target with less risk to human life (civilian and soldier) than the alternative, which would likely be an airstrike or significant application of suppressive fire.

Assuming a constant supply of energy, autonomous weapon systems are simply more effective at maintaining a constant defence because they do not suffer from fatigue, distraction or boredom. Furthermore, these systems do not have a self-preservation instinct and thus are less likely to overreact to a non-lethal threat than a human border guard. That is not to say that AWS are infallible. Chapter 9 contains an in-depth engagement with the potential negative consequences of deploying autonomous systems in these roles. Furthermore, increasingly autonomous systems remain vulnerable to cyber warfare or enemy interference, either at the programming stage or within their data-interpretation processes, a serious flaw given the damage that could be done by even a small alteration by a hostile actor. Overall though it appears that current technology would enable a weapon platform to exert autonomous control over engaging a designated target, following positive identification by a human.

5.3.2 'Software' – exploring emerging operational concepts

It is difficult to evaluate overall progress towards a final operational concept for an innovation that is capability-based, as opposed to a distinct platform, and still in development. However, the most widespread operational concepts in terms of concrete development efforts in state military doctrinal documentation, to date, centre on human–machine teaming (HMT). HMT is a broadly applicable series of concepts that focus on semi-autonomous and supervised autonomous weapons systems, as well as incorporating autonomous military technology into existing platforms. By definition a human-retentive approach, HMT presents fewer ethical, legal and technological challenges; however, this comes at the cost of willingly sacrificing some of the disruptive potential of fully autonomous systems. The underlying principle of human–machine combat teaming is capitalising on AMT to improve the lethality, survivability and/or utility of otherwise conventional human-centric combat units.

5.3.2.A Human–machine teaming – integrating semi-autonomous and supervised weapon systems

Adopting a HMT approach has substantive benefits, especially when paired with current (and expected) technology. Incorporating autonomous capabilities into modified existing platforms and operational structures would substantively be more practical for Singapore and Indonesia than attempting to adopt fully autonomous weapon platforms. HMT also capitalises on evidence that the public is more willing to accept the deployment of autonomous weapon systems to protect their own soldiers (Horowitz, 2016). The adoption of HMT operational concepts reflects Kasparov's observation that:

> *Weak human + machine + better process was superior to a strong computer alone and, more remarkably, superior to a strong human + machine + inferior process ... Human strategic guidance combined with the tactical acuity of a computer was overwhelming.* (Development, Concepts and Doctrine Centre, 2018)

Although Kasparov was referring to a 2005 chess tournament, it is telling that this quote appears in Joint Concept Note 1/18, written by the Development, Concepts and Doctrine Centre. At the heart of HMT is the recognition that computers, especially those with an autonomous learning capability, are better than humans at certain activities but inferior at others. These superior capabilities, when paired with humans, mean that autonomous systems can be used to achieve operational benefits even with today's technology.

The first operational concept recognises the benefit of teaming autonomous systems with human supervisors for logistics both in and out of combat. Multiple states have expressed an interest in utilising onboard AI to manage the maintenance schedules of complex manned platforms. For example, enabling aircraft to autonomously conduct diagnostics, coordinate maintenance cycles and predict when future repairs are likely to be necessary (Department of Defense, 2019). Furthermore, taking their lead from civilian organisations, such as mining companies in Australia (Ryan, 2018), militaries have realised that land supply convoys can be cheaply converted to drive themselves along pre-planned supply routes. In addition to increased resource efficiencies, eliminating the need for human drivers lowers the risk of casualties among a military logistics train. From a tactical perspective, using autonomous vehicles to ferry supplies and ammunition to soldiers in combat reduces both the risk to soldiers and the amount of weight they have to carry into combat. The fact that the UGVs designed for this purpose (like the MUTT) are often armed would be an operational bonus for small combat units.

Another example would pair AWS with manned platforms that provide guidance but, crucially, not active supervision. This would be more effective in high-tempo combat situations where the human partner cannot spare the mental bandwidth to control their robotic support platform or where the delay from an active control/data link would critically compromise the unmanned platform's effectiveness. Under programmes like Loyal Wingman (Wassmuth and Blair, 2018), human pilots would enable the use of lethal force by their autonomous wingmen immediately before an engagement, allowing the AWS to independently participate in aerial combat. An extension of this concept would be for an otherwise manned platform to autonomously fly the aircraft along an adaptive route to a given objective. This would allow the pilot to rest or focus on other tasks, increasing the endurance of strike aircraft at a comparatively low cost. The US Air Force (USAF) and Navy's interest in Boeing's autonomous air refuelling aircraft reflects an interest in developing a similar capability for a discrete supervised system to conduct the in-air refuelling of aircraft, further increasing the endurance of their strike fighters.

One of the most promising HMT concepts being explicitly pursued by multiple state actors, including the United States, China, Australia and the United Kingdom, is the integration of AI into operational headquarters. Their intent is to improve the Command and Control (C2) capability of commanders, and accelerate their Observe, Orient, Decide and Act (OODA) process. Current military headquarters are typically static, vulnerable high-value targets, in large part due to the number of support personnel who effectively act as intelligent filters between the battlespace and the commander. Even in situations where the operational tempo remains comparatively low, the average human command staff is not well suited to efficiently analysing large quantities of data from multiple sources in real time. This vein of operational concept revolves around developing a 'virtual assistant' AI program that would leverage a 'cloud brain' to analyse incoming intelligence data in real time (Kania, 2019), providing commanders with prioritised information and accelerating their OODA loop. Bringing Lieutenant Siri into the headquarters (at the strategic, theatre and operational levels) would be an effective response to the increasing tyranny of scale confronting commanders in a data-rich battlespace.

There are also disadvantages that stem from the human-centric nature of semi-autonomous and supervised weapon systems deployed within HMT operational concepts. From a military effectiveness perspective, by tying their deployment of autonomous military technology to human team members, a state is wilfully abandoning some of the main advantages of AWS, including their ability to operate in higher-tempo combat situations than human soldiers, and their immunity to psychological strain.

From a deployment perspective, maintaining a human-centric approach has two detriments. The first is that human operators need to develop a high level of trust in autonomous platforms, especially in combat. Anyone who has been in a self-driving car (or even one with adaptive cruise control) will know the feeling of unease when the car heads towards a red light. Trust can only be built over time with the benefit of high-quality training. Some scholars have suggested using augmented or virtual reality-based training to acclimatise soldiers to fighting alongside autonomous platforms. Related to this issue, deploying first-generation AWS in support of humans in combat creates risk. Complex computer systems typically either crash spectacularly or unexpectedly demand operator intervention, usually without enough time for a human to reasonably notice and intervene. The 2007 Oerlikon GDF-005 friendly fire incident was a tragic example of this (Shactman, 2007). Even when they do not lead to an 'unintended engagement',[4] critical failures of supervised autonomous systems would endanger their human operators and damage the trust that is so vital to the success of human–machine teaming.

5.3.2.B   Embracing the robotic warrior: Operational concepts for deploying LAWS

Although pairing autonomous weapon systems with human operators appears to have significant advantages, it would be folly to assume that militaries will always retain human involvement. This is simply because operational LAWS would outperform their human-supervised equivalents, at a lower economic and political cost, especially in combat operations with a higher tempo than humans can physically maintain. Morris summarises the issue:

> When robots with OODA loops of nanoseconds start killing humans with OODA loops of milliseconds, there will be no more debate. (Ryan, 2018)

This is particularly apparent when considering the deployment of AWS in an air superiority role. This would require the AWS have the capacity to operate autonomously, as the delay inherent in relying on a data link to convey instructions, further delayed by human reaction times, would cause them to be easily destroyed by foes that do not rely on relayed instructions. While current technology does not support deploying LAWS into unstructured combat environment with sufficient reliability that they consistently defeat human opponents, this has not stopped scholars and military planners from theorising.

The first proposal centres on the military advantage that could be gained from installing autonomous robotic pilots in outdated aircraft (Ackerman and Silver, 2016). Whether the robots would consistently defeat human pilots is debated (Finley, 2016), but irrelevant, given the sheer numbers of relatively effective combat aircraft this approach would allow a state to deploy. This is directly related to the popular conception of 'swarms' using AI to interpret general guiding principles. There is value in deploying a self-guiding swarm of cheap unmanned aircraft to disrupt airfield operations, harass or attack combat units, destroy material targets or provide near-constant surveillance (Bunker, 2015). Multiple actors are developing autonomous 'motherships' that include the capability to refuel UAVs and even 3D print replacements, increasing the endurance of a deployed swarm. Even if unmanned aircraft are not intended to be used for lethal force, we have already seen that individual soldiers are willing to adapt ostensibly non-lethal robotics to fulfil combat requirements, even if that entails duct-taping plastic explosive to the front (Lester, 2018).

The second scenario would be a direct but non-nuclear military confrontation between the United States and a near-peer military, especially if that peer has sophisticated A2AD capabilities or possesses autonomous weapon systems, for example, China. This scenario is reflected in US military documentation referring to a need for unmanned systems that can operate in denied environments. The third operational concept to consider here is cyberwarfare. While not a focus of this book, the benefits of weapons that can act without human authorisation are stark given the incredible operational tempo of outright cyberwarfare. In this regard, the focus on the Lethal component of LAWS is somewhat diverting attention from the fact that the recognition of the utility of autonomous systems in cyberwarfare is apparent in strategy documents across every state developer referenced in this chapter.

Finally, autonomous weapon systems could be utilised to strike a target that is located within a denied or hostile environment, which is too dangerous (or politically sensitive) for human combatants. Such an environment would not support the deployment of a UAV that is remote-operated via datalink, conventional air/ missile strike or the deployment of a special forces team, the three most common targeted killing methods. In this scenario, a LAWS that is pre-taught to recognise its target, based on a combination of biometric data, facial recognition and electronic signature (from a mobile phone, e.g.), could be despatched. The development of stealthy, fast autonomous UCAVs (like the Taranis and X-47B) is a clear response to this operational concept.

It would appear that the development of operational praxes for LAWS has reached a comparable point to armoured warfare during the interwar period. While there are clear themes emerging among conflicting concepts of how to integrate AWS into state arsenals, it has not yet become clear which strategic approach (or combination thereof) will become widely adopted in each domain as we move closer to a demonstration point. However, the common emphasis among developing states on human–machine teaming and swarming unmanned platforms is indicative that LAWS developers are influencing each other.

5.4 Level of recognition of LAWS as a disruptive military innovation among hegemonic competitors

The US defence establishment has clearly indicated an interest in pursuing increasingly autonomous military technology as part of a strategy to offset the rising strength of its competitors (alone or in alliance). This is not the first time that the United States has reflexively implemented an *offset strategy* in response to the challenge of a rival military. Prior offset strategies capitalised on major military innovations to disrupt and overcome the conventional (first) and nuclear (second) superiority of the Soviet Union. Contrastingly, the Third Offset Strategy reflects the dual-use nature of AMT and its low proliferation barriers. Instead of a single peer military gaining an advantage, in this case the United States fears that losing the race to develop and deploy AWS will allow near-peer militaries to subvert and disrupt its conventional military strengths, undermining the power projection that is essential to its hegemony. The Third Offset Strategy is therefore focused on encouraging the US military to rapidly innovate, *failing fast* along-side civilian partners in an effort to innovate, adopt and integrate increasingly autonomous military technologies, with an additional emphasis on cyberwarfare (Ellman, Samp and Coll, 2017).

Although the Third Offset Strategy was less visible in official documents in the first two years of the Trump presidency, the government confirmed its commitment to securing a lead in AI in July 2018 (Harwell, 2018). This commitment was reinforced over the subsequent year by its inclusion in the 2019 National Defense Authorization Act (Cronk, 2018),the singing of an Executive Order (Baker, 2019) and the release of a DoD AI Strategy. The latter demonstrated a renewed level of recognition of the dangers of failing to adopt increasingly autonomous systems and ceding initiative in related technologies to rival states. The strategy was also clearly influenced by the Third Offset Strategy. It primarily points to the benefits of incorporating AI for reducing risks to soldiers, improving resource efficiencies and shifting human personnel to focus on strategic decision-making, rather than dirty, dull or dangerous taskings (Department of Defense, 2019). More controversially this strategy made the claim that incorporating AI would improve implementation of international humanitarian law and reduce civilian casualties, claims that have been strongly questioned by various scholars and non-governmental organisations, such as the Campaign to Stop Killer Robots and Noel Sharkey.

It is also clear that the Chinese military is gaining on the United States in operational capacity and strategic reach (Allison, 2017). Despite the lack of explicit formal military doctrine (at least publicly), it is becoming increasingly clear, not least from public statements by senior Chinese leaders and defence scholars (Kania, 2019), that China believes that a new revolution in military affairs is beginning and that they do not want to risk being left behind again. Instead the People's Liberation Army (PLA) is aiming to capitalise on AI to improve their command decision-making and military performance (Work and Grant, 2019). Indeed, the Third Offset Strategy appears to have had a greater initial impact on Chinese policymakers than those in the United States, a reaction that was further reinforced by the 2016 success of AlphaGo, which Chinese officials saw as a 'sputnik-moment' (Work and Grant, 2019). It is unsurprising, therefore, that Chinese policymakers and military research organisations 'routinely' translate, analyse and cite scholarly and policy research published by their western counterparts on this topic (Allen, 2019).

More broadly, Chinese military development doctrine enshrines the importance of gaining superiority in 'domains of emerging military rivalry' (Raska, undated). For example, the director of the Central Military Commission's Science and Technology Commission stated that 'if you don't disrupt you'll be disrupted' (Horowitz et al., 2018). In some respects, China derives a level of advantage from being the rising challenger state; Chinese military expansion and modernisation is guided by the recognition that a conflict will likely turn on the PLA's capacity to counter and minimise the traditional power projection superiority of the United States. LAWS are seen as a pathway for overtaking US military power in the Asia-Pacific region, an approach that has been called a 'leapfrog strategy' and reflects a view that the character of warfare is changing to an 'intelligentized (智能化)' paradigm (Kania, 2017).

5.5 Comparative capacity of the United States and China to meet the development and operationalisation requirements of LAWS

5.5.1 Level of resource commitment to AWS-related research and development

The United States has secured a prominent role in the initial development of unmanned and autonomous systems, reflecting its status as the most well-funded military in the world (Tian et al., 2018). The sheer level of research and development funding has previously dwarfed that of competitor states. The US Department of Defense invested US$149 million in the autonomous technology priority area in 2015, followed by an additional US$18 billion across the 2016–20 period (Boulanin and Verbrugen, 2017). More recently, the 2019 DoD budget allocated a US$9.6 billion to programmes related to unmanned and autonomous systems (Klein, 2018), and the 2020 DoD budget request included a significant allocation for research related to autonomous systems (US$3.7 billion) and AI (US$927 billion) (Hourihan, 2019). It is particularly interesting to note that the US Army is funding over 60% of unique, cross-domain autonomous systems research projects in 2019, whereas its investment in UGVs in 2016–17 only amounted to just over 10% of what the USAF concurrently spent on UAV research (Gettinger, 2016). From a purely monetary perspective, it is apparent that the United States has recognised the importance of autonomous military technology to the effectiveness of their future military.

China's commitment to becoming a leader in AI research and development was made explicit in the 2017 New Generation AI Development Plan, but was a clear priority in the Made in China 2025 policy (2015) (Carter, 2018). While exact nationwide investment figures are not publicly visible (Boulanin and Verbruggen, 2017), it is apparent that China is heavily investing in this area. Annual Chinese defence spending is believed to have risen 620% in real terms between 1996 and 2015 (Work and Grant, 2019), and effectively tripled between 2007 and 2017 (Allen, 2019). Importantly, this resource investment has been guided by an explicit recognition that modernisation and replacement of legacy platforms are vital for future success (Allen, 2019). It is worth noting that the central government is not the only source of funding for research efforts in this field. For example, the cities of Xiangtan and Tianjin invested over US$7 billion between them (Carter et al., 2018). The Beijing municipal government has announced plans to invest US$2.12 billion in an AI development park (Horowitz et al., 2018) and a next-generation innovation fund that is expected to reach funding levels around US$14.86 billion (Kania, 2019). While China is orchestrating a strong campaign of investment in developing autonomous military technology, its efforts remain vulnerable to being undermined by a weakening economy and its hierarchical, top-down structure would have a cooling effect on the experimentation necessary for disruptive innovation (Kania, 2019).

5.5.2 National security innovation base

The United States has been able to draw on a world-leading National Security Innovation Base that includes many of the most influential civilian researchers, companies and military contractors to aid in its efforts to develop autonomous weapon systems. As the primary US defence research body, the Defense Advanced Research Projects Agency (DARPA) receives the lion's share of military research funding (29%) (Boulanin and Verbruggen, 2017), and was allocated an additional US$2 billion for AI research and development as part of the 'AI Next' strategy (Cornillie, 2018). In addition, DARPA runs competitions for civilian researchers and engineers with large cash prizes, which has proven an effective way to encourage innovation and participation by a variety of actors, and has been copied by other states (including Russia, China and the United Kingdom). Their interest in AI and autonomous systems has filtered into the service branch research organisations, which focus on shorter-term research projects that have the potential to impact the battlefield directly. For example, the US Army's *Robotic and Autonomous Systems Strategy* identified five 'capability objectives' to guide their integration of autonomous military technology (in combat and non-combat roles); while the US Navy maintains a focus on AI across all six of its Integrated Research Portfolios; and the US Air Force's *Unmanned Aircraft Systems Flight Plan 2016–2036* focused on mini-UAVs (Otto, 2016).

Roughly coinciding with the signing of the *Executive Order on Maintaining American Leadership in Artificial Intelligence* (Baker, 2019), the establishment of the Joint Artificial Intelligence Center (JAIC) was one of the centrepieces of the DoD AI Strategy. The JAIC was allocated US$1.7 billion to support its initial establishment (Cornillie, 2018). The initial purpose of the JAIC was to provide the 'critical mass of expertise' needed to rapidly identify, prioritise and operationalise AI research efforts across the DoD. Where DARPA focuses on long-term projects, the JAIC is closer to the service branch laboratories in its focus on short-term AI-enabled projects that can be rapidly developed into capabilities for warfighters (Cronk, 2018). The JAIC also plays a role in developing an integrated library of shared tools for AI research and acts as the main linkage builder between DoD and civilian experts. The latter line of effort is supported by the Defense Innovation Unit (DIU), effectively a physical DoD outpost in Silicon Valley, which was intended to encourage start-up-led rapid defence innovation (Mehta, 2018). Considered effective, at least by near-peer competitors (Russia and China developed similar offices), the DIU was subsequently allocated US$139 million of core funding and US$25 million for university partnerships under the 2020 DoD budget request (Hourihan, 2019).

By comparison, the common focus on intellectual property theft and cyber espionage belies the fact that China is also developing a surprisingly advanced national security innovation base, which has a particular focus on emerging technologies. Recent developments include committing to spending 2.5% of GDP on research and development as part of the *13th Five Year Plan of 2016–2020 'Internet Plus'* and building the world's fastest supercomputer in 2016 (Brown and Singh, 2018). Propelled by a rapid expansion and diversification of funding, with access to the second-largest talent pool (Allen, 2019), Chinese domestic research institutions are making significant contributions to the development of increasingly autonomous unmanned platforms. For example, the National University of Defence Technology opened two research centres in the past year that are focused on unmanned systems and AI (Horowitz et al., 2018). Furthermore, China has already overtaken the United States in several metrics of innovation in this space, including filing almost twice as many relevant patents and securing the same number of places as the United States in engineering university rankings as far back as 2015 (Allison, 2017). Overall it is becoming increasingly clear that the view of China as limited to emulating and stealing technology because it lacks the capacity to compete with western powers as innovators at the forefront of emerging military technologies is not fully reflective of the facts (Brown and Singh, 2018).

5.5.3 Civilian participation

Following the Third Offset Strategy's approach to innovation, the US Department of Defense operates in partnership with non-military research bodies and has historically been a major funder of civilian research, both commercially and within universities. The DoD is providing additional research funding to attract talented researchers into 'strong and stable ... clustered' research partnerships with DoD. The DoD AI Strategy further reflected a commitment to promoting research partnerships with the 'open-source community', multinational firms and international partners to advance emerging technologies (Department of Defense, 2019). However, the US civilian research sector has demonstrated a marked reluctance to participate in military research, especially when it is directly related to weapon systems (Singer, 2009). Only a limited number of prominent civilian universities have accepted military funding for research related to AI and unmanned platforms. These include Carnegie Mellon University, which has its own robotics programme and cooperates with both DARPA and the Army Research Laboratory; and the Caltech Center for Autonomous Systems, which has a particular focus on developing autonomous systems for disaster relief and public safety. The majority of related research is, therefore, being conducted without military funding and is intended for civilian applications. An increasingly significant amount of related research is even being funded by private firms. While corporate developers including Apple, Alphabet and Boston Dynamics have all made substantial advances in related research, the withdrawal of Google from Project Maven demonstrated that private firms are not immune to this resistance among their staff (Cornillie, 2018).

China's active 'innovation driven' civil–military fusion has been notably more successful than the United States, partially due to the personal commitment of Xi Jinping (Kania, 2019). Large technology firms such as Ali Baba Cloud and Baidu are participating in Chinese government-funded national research laboratories, and even became part of a 'national team' developing AI (Horowitz et al., 2018). Commercial firms have been joined by ostensibly civilian university researchers in this effort, with several reports identifying the Harbin Institute of Technology, the North China University of Technology and Tsinghua University as having significant research partnerships with the PLA. For example, over the past two years the latter (which is regularly referred to as the Chinese MIT) partnered with the Central Military Commission to launch the *Military–Civil Fusion National Defense Peak Technologies Laboratory* and the *High-End Laboratory for Military (Artificial) Intelligence* (Kania, 2019).[5] China is clearly prioritising military–civil fusion with the goal of becoming the global leader in AI by 2030 (Allen, 2019). However, it is important to reiterate that the higher level of control exercised by the Chinese government over domestic technology companies will limit their ability to pursue the sort of risky, experimental innovation that leads to revolutionary, as opposed to incremental, advances (Kania, 2018).

5.5.4 Challenges developing and securing top-level talent relevant expertise

Despite this level of resource commitment, both states are struggling to attract and retain top-level talent in relevant fields, such as AI and robotics. The United States recognised this challenge in the DoD AI Strategy. One of the priority areas noted in the strategy was targeted recruiting of world-class researchers, supported by a renewed emphasis on multilateral research partnerships and investment in younger researchers (Department of Defense, 2019). Interestingly, the DoD flagged an interest in building its organic AI skill base through providing 'curated training programs' to in-service DoD civilians and military personnel. Supported through non-traditional recruitment and secondment of civilian experts, this training would build relevant skills and interest across all levels of the US military (Department of Defense, 2019). While this is a novel approach, the participation of high-profile civilian researchers and engineers in NGOs opposed to the military application of autonomous systems suggests that it will be difficult to attract the necessary interest from the admittedly deep pool of related expertise in the United States.

China also possesses a rapidly expanding STEM-qualified workforce, which it draws from universities that graduate four times as many STEM undergraduates and more STEM-related PhDs than their US counterparts (Allison, 2017). This advantage is further reinforced by the fact that Chinese nationals compose 25% of all US university graduates in STEM fields (Brown and Singh, 2018). Furthermore, China continues to build connections with innovative foreign firms, with Chinese investment evident in 16% of all venture capital investments in US-based start-ups between 2015 and 2017 (Brown and Singh, 2018). However, China lacks comparable access to the most skilled and experimental research-ers (Kania, 2019). This gap threatens China's capacity to innovate on the cut-ting edge, or even to integrate illicitly gained information into domestic research efforts. The government has clearly recognised this weakness and is actively pro-moting cooperative research with foreign universities in response (Allen, 2019).

5.5.5 Maintaining sufficient access to relevant data for training AI-enabled systems

The requirements for developing and operationalising autonomous platforms go beyond the number of scientists, dollars or programmes. One of the major issues with AI is that it needs to be 'taught' so to speak. With current machine learning techniques this requires masses of task-specific data, running hundreds of scenarios that the AI software can then learn from. For example, the Google AlphaStar AI that defeated StarCraft II professional gamers was in fact a series of AI agents, which were initially 'trained' using the data from professional replays before competing in iterative tournaments against each other across the equivalent of up to 200 years of real-time gameplay. In order to gather the data necessary to its training, StarCraft II recently allowed players using its European servers to compete in ranked matches against AlphaStar agents for a limited time on an opt-in basis (Kan, 2019).[6] While the United States currently has an advantage in this space (stemming from the number of popular internet services, like Gmail, that are based in the United States) and superior access to high-level computing power, this advantage is decreasing. China is expected to become home to almost 30% of global data by 2030, a market share that will increase its ability to develop and train AI (Carter, 2018).

However, an important caveat is that the term 'relevant' is used quite specifi-cally here. 'Teaching' a system requires that the developer have access to signifi-cant amounts of task-specific data (Horowitz, 2019). For example, if a developer intended for an unmanned maritime vehicle to autonomously scan passing vessels and only fire on those identified as being from an opposing state, the enabling software would have to be trained with sensor data from those opposing vessels not friendly or civilian ships. While computer-generated or 'synthetic' data can be used to reduce this requirement (Allen, 2019), it is only a stopgap and does not fully replace the need for high-quality relevant data when 'training' AI.

Overall it would be unwise to assume that there is some kind of permanent barrier to China gaining a technological advantage in this space, even purely from a comparative resource standpoint. Rather Chinese research efforts and partner-ships are already producing world-class technologies even, on occasion, ahead of the United States. While China still lags behind the United States in significant areas and their doctrinal approach to LAWS is not fully developed, it is no longer feasible to assume that the United States is necessarily guaranteed to secure a relative advantage following the emergence of LAWS.

5.6 Other state developers of AWS active in Southeast Asia

5.6.1 Russia

Russian research and development efforts in AMT are less well reported in the western media than US or Chinese developments. The lack of media attention belies the fact that Russia has identified military robotics as a priority research area, committing approximately US$346 billion across 2016–25 (Boulanin and Verbruggen, 2017). The Russian Foundation for Advanced Studies (FPI) fills a similar role to DARPA in this effort. The FPI controls an annual research budget of US$78 million (Cooper, 2016), which is allocated along the lines of its five priority 'directions' (Направления), one of which is the National Centre for the Development of Technologies and Basic Elements of Robotics.

To an even greater extent than their Chinese and American counterparts, the FPI emphasises long-term partnerships with other research institutions. An interesting example is the Laboratory of 'Intelligent Constructions' (Интеллектуальных ко нструкций), which operates in conjunction with the Institute for the Development of Research, Development and Technology Transfer. Another FPI laboratory is the Laboratory for the Development of Optical Devices of the New Generation, which is a collaboration project with staff from 15 Russian (and one Japanese) universities that is using quantum mechanics, plasmonics[7] and advanced lidar to develop next-generation sensors that are sensitive to the molecular level. While the official descriptions provided by FPI of each of their programmes emphasises their domestic impacts, this research has clear AWS applications. To increase engagement with researchers who are not partners of FPI laboratories, the FPI also holds innovation competitions, styled on DARPA's approach. For example, in 2018 a design tournament was held in Vladivostok, which featured 26 Russian universities and a defence contractor, competing with remote-operated and autonomous UMVs.

Where its domestic research or production capacity falls short, Russia has previously purchased platforms from other states. For example, Russia purchased Israeli-made UAVs, while developing a modern domestic model (it now has the KT Orion and Zala 421-20). This complements a powerful and innovative defence industry that is actively developing AMT. For example, MiG Corporation and Sukhoi have disclosed proposals for unmanned next-generation strike aircraft. The Sukhoi Su-57 would allegedly be able to autonomously 'decide exactly what type of arms and ammo it needs' and operate at significantly higher speeds than would be safe for a human pilot (Sputnik News, 2017). The Su-57 was initially deployed in Syria in February 2018 and is being used as a technology testbed for in-development sixth-generation autonomous strike aircraft (TASS, 2018). Of the limited public information available, it appears that the Russian military is also pursuing autonomy for ground and maritime vehicles. The most advanced Russian UGCV is the Uran-9. While officially a remote-operated 'drone', the Uran-9 is closer to a supervised autonomous weapon system with the ability to autonomously manoeuvre, and identify and acquire targets, although firing requires authorisation from the human operator. In May 2018, the Russian Deputy Minister of Defence confirmed that the Uran-9 had been deployed in Syria, although whether it engaged in combat is unconfirmed (RT, 2018). This is the first time a state has admitted to deploying an autonomous ground combat vehicle into an active combat zone.

The Russian Federation's efforts to develop increasingly autonomous weapon systems can be seen largely as an effort to gain a greater level of military parity with the United States and Europe. This position seems to have been summarised by Putin's statement that 'the one who becomes the leader in [autonomy] will be the ruler of the world', which was followed by a less commonly cited acknowledgement that 'it would be strongly undesirable if someone wins a monopolist position'. Russia has been opposed to a development ban during meetings of the intergovernmental group of experts on LAWS but has not actively blocked discussion.

5.6.2 United Kingdom

The United Kingdom has also emerged as a major state developer of AMT. The UK Ministry of Defence's (MOD) approach to integrating AMT into their force organisation is guided by its internal think tank, the Development, Concepts and Doctrine Centre (DCDC). A recent DCDC paper of interest is Joint Concept Note 1/18. Released by the DCDC in May 2018, this document seemed to signal a shift away from the prior UK MOD definition of autonomy (which required a device be capable of understanding higher-level intent and direction).[8] In JCN 1/18, the DCDC suggests that autonomous systems will develop across three stages of maturity. First, they will augment existing capabilities, requiring only a slight modification of existing operational concepts; eventually AWS will parallel the capabilities of their manned equivalents in key tasks; before they finally supersede manned platforms in key operational areas, rendering manned platforms obsolete. Examining the policy and strategic discussion emanating from the DCDC, it becomes clear that the UK MOD is focusing on human–machine teaming in a manner similar to its US counterpart.

This focus is also apparent in the direction of UK state research and investment. UK MOD research and development investment primarily occurs through the Defence Science and Technology Laboratory (DSTL), which is roughly equivalent (albeit significantly smaller) to DARPA. Relevant internal research focuses include the Future Sensing and Situational Awareness Programme, which is developing sensor systems that can operate in denied environments; and the Future Threat Understanding and Disruption Programme, which is specifically intended to identify and develop responses to emerging technologies or capabilities to lessen the effect of 'future shock' on the British military. As part of its broad responsibility for innovation across the British military, the DTSL funds external researchers in the university and commercial sectors, and even purchases technology off the shelf where it has a military application. DTSL holds similar competitions to its American cousin, although more frequently at a smaller scale. However, efforts to encourage civilian participation in autonomous weapon-related research have been hampered by opposition from sections of the scholarly community. UK and European Union researchers are disproportionately represented in groups opposed to the development of LAWS, including Researchers for Peace.

Although the United Kingdom has expressed support for some form of regulation being placed on LAWS, it has not supported a developmental ban. This has allowed the defence industry in the United Kingdom to participate in several high-profile development projects. These include the EuroSWARM, which is a distributed swarm logic-based system for deploying cheap unmanned robots in tactical surveillance roles; the Taranis UCAV, which can conduct fully autonomous strike missions, although BAE has ostensibly only produced a demonstrator model; and the European Future Air Combat System, which incorporates human–machine teaming to improve the performance of manned fighter aircraft. All of which are smaller in scope than Ocean 2020, which aims to develop unmanned platforms for integration with manned combat vessels. Overall it is clear that the United Kingdom is embracing autonomous technology within the bounds of a declared repudiation of what it considers to be fully autonomous weapon systems.

5.6.3 Israel

Israel certainly is not new to the realm of military robotics. They were among the first states to effectively use unmanned aircraft in combat and are a leading exporter of UCVs. Israel's decision to invest in robotics and unmanned technology is understandable given their small population, the hostility of its neighbours and the ongoing civil war in neighbouring Syria. Further, exporting military robotics has proven both lucrative and influence winning for Israel. Remote-operated and semi-autonomous weapon systems developed by Israel have turned up in several states, including Iraq and Jordan (Ewers et al., 2017). Unfortunately, very little specific information about Israeli military research and development funding is publicly available (Boulanin and Verbruggen, 2017).

Israel has made progress in developing autonomous capability in all three theatres. Autonomous ground vehicles like the IMI Systems AMSTAFF and the Guardium are available to purchase, as are autonomous sentry turrets like the Samson series from Rafael. These are complemented by a wide range of underwater and surface vessels and an expansive series of UAVs with varying levels of autonomy. Overall, Israel is developing autonomous technology at a remarkable rate, subsidised by ongoing sales of increasingly autonomous systems to state and non-state actors.

5.6.4 Republic of Korea

As a highly technologically advanced society with highly influential research institutions and a burgeoning industrial robotics market, it should not be surprising that the Republic of Korea (ROK) is one of the leading developers of autonomous military technology. The ROK government has invested heavily in the development of robotics (US$840 million in the 2016–2020 period) and planned to put a robot 'in every household' by 2020. This funding has spurred the participation of South Korean Chaebols (conglomerates) and universities, particularly KAIST and POSTECH.

The ROK state organisations responsible for encouraging defence research reflect a shift in domestic military thought that occurred in 2005 (Moon and Lee, 2008). Marked by the release of the Defense Reform 2020 Plan, the South Korean government has been actively working to increase the sophistication of its domestic defence industry (Moon and Lee, 2008). The main ROK organisations responsible for encouraging military innovation, research and development are the Defense Acquisition Programme Administration, which is mainly responsible for managing the acquisition of new weapon systems and technology from domestic and international suppliers; the Agency for Defense Development, the main research arm of the Ministry of Defense, operating approximately 56 major research laboratories; and the Korea Institute for Defense Analyses, which is an external analysis body that provides expert advice to the Ministry of Defense on what innovations best fit the current needs of the ROK military. These state bodies are collectively responsible for investing the military's resources in research, development and procurement.

The other side of the Defence Reform 2020 Plan is the South Korean defence industry. South Korea was ranked the 12th largest global arms exporter in 2017 (Wezeman et al., 2018). The export market for ROK weapons rose from US$253 million in 2006 to US$3.19 billion in 2017, an increase of 1160.86% (Grevatt, 2018). Most large ROK defence companies are component parts of Chaebols, reflecting the dual-use nature of autonomous technology. For example, two of Hyundai's subsidiary companies build military systems (Hanwha Land Systems, 2018), while another section is developing autonomous civilian cars (Edelstein, 2018). South Korean companies are generally less circumspect than their American counterparts when describing the autonomous capabilities of their systems. For example, a senior DoDaam engineer admitted to the BBC that they had designed the Super Aegis II as fully autonomous and only added a human supervisory mode after a customer expressed concern (Parkins, 2015).

The final South Korean research actor is civilian universities. While there is military-related research occurring at almost every university in South Korea, it is worth focusing here on Korea's premier engineering and science university, KAIST. By design, the state-run KAIST is situated in Daejeon, the heart of the Korean defence industry, and is heavily involved in military research. Consider the Unmanned Systems Research Group (USRG), which is a single laboratory in the College of Aerospace Engineering. USRG researchers have developed a human-scale robot that can effectively fly an unmodified fighter jet in simulated combat conditions; they are currently developing deep learning-based object avoidance software and an AI model for fully autonomous capable UCAVs (Unmanned Systems Research Group, 2018). The majority of senior researchers at KAIST have completed their compulsory military service, which typically involves at least one deployment to the de-militarised zone and are generally more aware of the realities of combat conditions than the average engineer.

5.7 Conclusion: moving towards a demonstration point

Considering the status of LAWS as a disruptive military innovation through the lens of current technology, it appears that the 'hardware' component of LAWS has not sufficiently matured. Even with the massive resource investment, frontline combat robots would continue to struggle in a dynamic ground combat environment. However, it is also clear, even from publicly available data, that the rate of technological development is rapidly bringing that point closer. The main factor will be related to improving the reliability with which machines adapt to unexpected conditions in a combat setting.

The development of LAWS operational concepts is also clearly underway. To date there has been a clear preference for incorporating AI and autonomous weapon systems into a human-centric conception of warfare. Improving the efficiency and effectiveness of the OODA loop of human commanders will be vital as the operational tempo and complexity of warfare continues to increase. It will be interesting to see whether states continue to focus on the development of doctrine that preserves traditional operational structures and remains human-centric.

In concluding this chapter, it appears likely that only large, wealthy states will have the infrastructure and resources to initially acquire and effectively deploy full LAWS. However, this should not constrain a scholar from examining the impact of less sophisticated semi- and supervised autonomous weapon platforms and the spread of related technology in the aftermath of a LAWS demonstration point. Indeed, the emerging consensus among academic, industrial and policy literatures increasingly holds that, in the absence of a pre-emptive and effective development ban, autonomous weapon systems will mature and begin to proliferate. A study conducted by the US Joint Forces Command estimated that the LAWS demonstration point could arrive by 2025.[9] Which response is taken by leading ASEAN states will be largely determined by their individual security priorities and resource availability; and is the focus of the following chapters.

Notes

1 Artificial Intelligence can be described as 'the use of computing power, in the form of algorithms, to conduct tasks that previously required human intelligence' (Horowitz, 2019).
2 'A fully autonomous Lethal Autonomous Weapon System (LAWS) is a weapon delivery platform that is able to independently analyse its environment and make an active decision whether to fire without human supervision or guidance' (Wyatt and Galliott, 2018).
3 It was confirmed to the author by a former Australian Army officer that the Saab Giraffe AMB can detect projectiles as small as a 7.62-mm rifle round.

4   The use of force resulting in damage to persons or objects that human operators did not intend to be the targets of U.S. military operations, including unacceptable levels of collateral damage beyond those consistent with the law of war, ROE, and commander's intent.

This is the US Department of Defense's term for this sort of incident.

5   Dr Elsa Kania's 2019 testimony to the congressional United States–China Economic and Security Review Commission contains a highly detailed review of Chinese military–civil fusion projects.

6   Interestingly, following this initial opt-in, players would encounter AlphaStar opponents anonymously through the normal matchmaking process. Obscuring their identity from players was intended to ensure that the AI agents were able to train in 'realistic' game conditions.

7   The study of the interaction between electromagnetic field and free electrons in metal.

8   This definition was originally presented in Joint Doctrine Note 2/11: The UK Approach To Unmanned Aircraft Systems. This was replaced in August 2017 by Joint Doctrine Publication 0-30.2: Unmanned Aircraft Systems, which retained the same definition.

9   This study is no longer publicly accessible, a public citation of the study can be found in Krishnan (2009).

6 Evaluating Indonesia's adoption capacity

6.1 Introduction

Indonesia has emerged as a leading economic and military power among the Association of South-East Asian Nations (ASEAN) member states. Setting aside the contention that Revolutions in Military Affairs have historically been the province of great powers and their rising competitors, this chapter analyses the extent to which Indonesia as a small but influential regional power is likely to become a secondary adopter, whose reaction to the emergence of LAWS will have a meaningful and disruptive effect on the hegemonic tension between China and the United States as well as on the future of security in the Asia-Pacific.

This chapter will evaluate Indonesia's adoption capacity, which is based on five diffusion variables and can be used to evaluate which response option would be most effective for Indonesia following a LAWS demonstration point (Horowitz, 2010). The first diffusion variable is the security-threat environment, the influence of traditional and non-traditional security threats on doctrinal and procurement decisions. The second variable is resource capacity, which includes military expenditure, the sophistication of Indonesia's domestic military-industrial base, foreign arms acquisition and the national security innovation base. The third variable, Indonesia's Organisational Capital Capacity, has three sub-variables: Critical Task Focus, Level of investment in Experimentation and Organisational Age. The final two diffusion variables are the receptiveness of domestic audience towards autonomous military technology and the Indonesian military's ability to develop or emulate a specialised operational praxis to effectively deploy the disruptive innovation.

There has been a recent push by Indonesian policymakers for the archipelago state to be recognised as a rising regional great power, a nationalistic push for prestige and influence in a region expected to rise in importance. Statistically, this push has some merit, since by 2017 Indonesia had reached a four-year GDP growth rate peak of 5.2% (International Institute for Strategic Studies, 2018) to become the largest economy in Southeast Asia (The World Bank, 2017), with a level of military expenditure second only to Singapore among ASEAN member states (International Institute for Strategic Studies, 2018). Buoyed by this growth, Indonesia has pursued a foreign policy of regional independence from China and

DOI: 10.4324/9781003172987-6

the United States, known as 'Pragmatic Equidistance' (Laksmana, 2017). Unlike other rising powers, particularly in the BRICS block, Pragmatic Equidistance is based in strategic positioning, non-confrontation and soft-revisionism, favouring the assumption of the role of a trusted interlocutor rather than that of an overt leader (Santikajaya, 2016). This distinction can be seen in Indonesia's influence within ASEAN as well as its self-appointed role as a 'neutral intermediary' in the South China Sea territorial disputes between ASEAN member states and China.

Indonesia's push for regional influence has, however, been repeatedly undermined by its own domestic pressures and instability. Despite a notable nationalistic streak (Wijaya, 2018), regular separatist movements, insurgencies and organised criminal groups have undermined Indonesian sovereignty and security, leading to embarrassing ASEAN resolutions, arms embargoes (Laksmana, 2014) and even a UN-supported military intervention led by a neighbouring state (Blaxland, 2015). For Indonesia, becoming an early adopter of Lethal Autonomous Weapon Systems would be a powerful but symbolic move, putting it on the forefront, at least within Southeast Asia, of a new state power paradigm.

This chapter will demonstrate that, despite its recent economic and military growth, Indonesia lacks the capacity to successfully become an early adopter of LAWS. Rather it would be better served by attempting only limited adoption within a response that primarily relies on diplomatic rebalancing. By doing so this chapter will inform a broader understanding of how ASEAN states are likely to respond to the emergence of this paradigm-shifting major military innovation, which is central to the core research question of this book.

6.2 Evaluating Indonesia's adoption capacity

6.2.1 Security threat environment

By any measure Indonesia is not lacking in security threats. In addition to traditional state-based threats to its territorial integrity, Indonesian security forces have had to respond to internal instability and multiple insurrections, while being limited by the ongoing pressure of non-traditional threats (such as human trafficking, piracy and terrorism). In the context of emerging autonomous weapon systems, the most relevant of these threats are piracy and organised crime, state aggression and territorial incursion, and internal rebellion and terrorism. These threats, as well as Indonesia's responses, should be considered within the context of the broader Southeast Asian security environment. This section offers an insight into how Indonesia's focus on internal security threats in the maritime and aerial domains will influence its approach to autonomous weapon systems.

While it is unlikely that a neighbouring state would invade Indonesia, it still faces threats to its sovereignty and territorial integrity. Despite its recent attempts to assume a leadership role in regional organisations (principally ASEAN and the Non-Aligned Movement), Indonesia has a history of conflict in the region that continues to influence its relations, particularly with Malaysia and Singapore. In the eyes of key military decision-makers, Malaysia is benchmarked as a 'military peer competitor', while tensions with Singapore intermittently flare into diplomatic incidents (Schreer, 2015). Indeed, Indonesia has been involved in recent territorial disputes with fellow ASEAN member states, including Malaysia, over incursions into its Exclusive Economic Zone (EEZ) (Schreer, 2015) and disagreements over ownership of islands in the Sulawesi Sea (Butcher, 2013). While intraregional tensions and disputes are an important factor in Indonesia's security environment, the pursuit of prestige through modernisation and their involvement in South China Sea territorial disputes have arguably been more significant factors driving the Indonesian military towards increasingly autonomous weapon systems.

Despite Indonesia's attempts to maintain neutrality by claiming not to be a party to the South China Sea disputes, Chinese encroachments have escalated recently. A key flashpoint in Indonesian–Chinese territorial disputes is the area near the Natuna Islands, which have escalated to include standoffs between government vessels and the demolition of captured fishing vessels (Parameswaran, 2017). Specific incidents include the Indonesian Navy firing on a Chinese fishing vessel in mid-2016, allegedly injuring a crewmember (Reuters, 2016), while in December 2017 a larger vessel, the Fu Yuan Yu 831, was boarded and seized (Parameswaran, 2017). While there has been an escalation in political rhetoric against illegal Chinese fishing, this has not translated into a broad agreement among Indonesian policymakers on how to respond (Syailendra, 2017). Despite the absence of a solidified political direction, the Indonesian Navy has significantly expanded its permanent presence in the area in an effort to deter intruding vessels and reassert Indonesia's claim (Parameswaran, 2019).

However, these ongoing territorial disputes have limited Indonesia's ability to maintain its preferred foreign relations approach of 'Dynamic Equilibrium', which prioritises secure, informed neutrality. A key component of this approach is the belief that maintaining regional stability and growth among ASEAN member states is crucial for preventing either great power from gaining too much influence in Southeast Asia. Limiting the influence of great powers allows Indonesia to act as a 'trusted interlocutor' between state actors (Syailendra, 2017), which in turn is part of Indonesia's perceived leadership role in ASEAN (Syailendra, 2017). Indonesia's role in ASEAN has been further complicated by the broader military build-up throughout Southeast Asia, which has generated fears of an impending regional arms race, exacerbating regional tensions.

Another security concern, which has historically received more attention from Indonesian defence planners, is the prospect of another province descending into rebellion or insurgency. The severity of this threat in the mind of Indonesian defence planners is apparent from the Indonesian military's continued commitment to 'territorial postings', which deploy TNI personnel alongside each level of government, even to the village level, to ensure internal security. Furthermore, Indonesia actively advocated for the creation of an ASEAN peacekeeping force in 2004 for deployment in support of member governments during internal conflicts and strife, although this was blocked by Singapore, Thailand and Vietnam (Nathan, 2006). Connected to this concern is the possibility that one of Indonesia's neighbours could support, fund or protect a breakaway faction or province, as happened with the Australian-led INTERFET into East Timor. Therefore, the TNI is understandably interested in the capacity of AWS to be deployed at comparatively low ongoing resource cost in long-term surveillance and border protection roles. Furthermore, the appeal of instruments of state violence that unquestionably follow commands in politically and ethically problematic internal conflicts to authoritarian and illiberal governments should not be underestimated. Finally, autonomous and remote-operated weapon platforms would have clear appeal to a military whose main strategic doctrine, the Total People's Defence System, centres on sustained guerrilla defence in the outer islands.

Indonesia also has endemic non-traditional security issues, including natural disasters, corruption, organised crime and piracy. Given the immense importance of the Straits of Malacca, which fall partially within Indonesian territory, piracy and maritime robbery are also key non-traditional security threats. It is not hard to see how Indonesian defence planners would be attracted by the potential offered by autonomous patrol boats or surveillance aircraft that could operate at a fraction of the running cost of a manned naval vessel (Mugg, Hawkins and Coyne, 2016). Indonesia's response to these security issues is of particular importance because the Sulu Sea also falls partially within its territory (Ray, 2018).

6.2.2 Resource capacity

As outlined above, Indonesia's economic strength has been notable even within the context of a region characterised by rapid economic growth in recent years. A concurrent growth in military spending can be attributed to rising nationalism, regional security concerns and the enduring influence of the TNI in domestic politics (Sebastian and Gindarsah, 2011). However, significant manpower and equipment maintenance costs and an enduring 'defence-commitment' gap between promised and delivered funds have diluted Indonesian military modernisation efforts.

The Indonesian defence budget reached 120 trillion Indonesian rupiah in 2017 (US$8.98 billion), which was the second highest in Southeast Asia (trailing Singapore). IHS Markit (2016) predicted that Indonesian defence spending would undergo the fifth-fastest global growth between 2016 and 2025. This contention was supported by the fact that spending rose by 122% between 2008 and 2018 (Tian et al., 2018). While the final allocation of funds was 11.8% lower than the original proposal, the 2018 state budget allocated substantially more funding to the Defence Ministry than its health and education counterparts (Chairil, 2018). Indonesia is not unusual in this regard; driven by regional distrust and China's increasingly belligerent territorial claims, the overall trend among ASEAN states has been to increase their defence spending at a regional average rate of 9% since 2009 (Laksmana, 2018).

While its top-line military spending is high, Indonesian defence spending is far below the global average as a percentage of GDP (0.8%) (Chairil, 2018). Considering Indonesia's economic peers, it is slightly more than Mexico (0.5%), yet substantially lower than Australia (2%), Turkey (2.2%) and the Republic of Korea (2.6%).¹ In the latter case, this difference equated to approximately US$30 billion in greater defence spending by the Republic of Korea in 2017 (Boulanin and Verbruggen, 2017). Despite repeated promises to raise defence spending as a percentage of GDP by civilian policymakers, there has been no concrete progress made towards achieving an allocation of even 1.5% of GDP, which was pledged in 2013. Until this changes, the potential for the TNI to develop autonomous military technology is limited.

Given that Indonesia remains a regional middle power, it is not surprising that its financial capacity is dwarfed by great powers, such as the United States, which devoted US$18 billion across the 2016–20 period solely to the development of autonomous military technology (Boulanin and Verbruggen, 2017). While the TNI does not have the resources to compete with existing great power developers, the Indonesian government could follow the Republic of Korea's example. Over the 2016–20 period, the ROK government committed to investing US$840 million directly into its domestic innovation base to encourage the development of autonomous military technology and robotics. Though Indonesia has a growing economic capacity, its government would need to commit a greater proportion of this capacity directly to military modernisation and supporting the domestic development of enabling technologies.

6.2.2.1 Domestic military-industrial base

A prominent component of Indonesia's modernisation efforts under the 'Minimum Essential Force' strategic doctrine is expanding the size and sophistication of its domestic military production capability (International Institute for Strategic Studies, 2018). Despite the renewed emphasis on building this sector, which was reflected in its designation as the third stage of the Minimum Essential Force concept, the modern Indonesian defence industry remains underdeveloped and largely focused on manufacturing less advanced arms. It is dominated by four state-owned firms: *PT Pindad, PT PAL Indonesia* and *PT Lundin*, and *PT Dirgantara Indonesia*,² who broadly focus on land, sea and air, respectively.

The establishment of the Defence Industry Policy Committee, or *Komite Kebijakan Industri Pertahanan* (KKIP), was arguably one of the more significant early responses (Haripin, 2016). Initially established in 2010, the KKIP was expanded by new defence industry regulations in 2012 and 2013. The modern KKIP, while still chaired by the Indonesian President, is governed by the Minister of Defence as its managing director (S. Rajaratnam School of International Studies, 2014). The Minister of Defence is then responsible for appointing an implementation team (comprised of bureaucrats and officials with defence industry expertise) and the expert team (which is recruited from public and private sector security researchers and experts) with responsibility for advising future decisions and policy direction (Haripin, 2016). The broader KKIP organisation is structured around six task-delineated divisions that reflect its broad responsibility for developing and coordinating strategic policy and plans for the defence industry.[3] In March 2018, the KKIP held a major press conference to promote the progress it had made, while downplaying the significant issues that remain in the security-defence industry (Parameswaran, 2018). Although the establishment of a directing committee demonstrates an encouraging level of commitment by the Indonesian government, the TNI has consistently reiterated that modernisation will require a greater resource allocation, a claim that was explicitly made in the 2015 Indonesian Defence White Paper.

Military modernisation efforts have been undermined by several factors, including chronic underinvestment in research and development (Chairil, 2018), economic conditions imposed by the International Monetary Fund after the 1997 Asian Financial Crisis (Sebastian and Gindarsah, 2011) and regulatory uncertainty (Chairil, 2018). One of the early KKIP initiatives prioritised improving the provision of the secure long-term funding needed to enable significant long-term investment in the major state-owned arms companies. PT Pindad received a cash injection from the state of approximately US$53 million in 2015 (Silviana and Danubrata, 2015) and, by 2018, had diversified into commercial manufacturing. Also in 2018 a series of agreements were finalised to secure favourable financial treatment for the security-defence industry, including easier access to credit and insurance. Recent efforts to improve the domestic arms industry have been prompted by the international prestige, income and influence gained by a state with a well-respected domestic arms industry, which would be beneficial in Indonesia's push for recognition as an emerging regional great power.

Another issue is that domestic arms manufacturers are competing against foreign companies that enjoy a dominant technological lead. This gap continues to increase as smaller state arm producers become constantly caught up developing a comparable product, while the leading firms have already shifted focus to a new capability (Bitzinger, 2017). The Indonesian response reflects the notably nationalistic streak present within recent defence decision-making, mandating the involvement of domestic suppliers and imposing a technology transfer requirement on high-value arms procurements. This could act as a barrier to purchasing autonomous weapon systems, which are likely to be under technology transfer restrictions similar to those imposed by the United States on UCAV sales.

Considering these problems with the domestic industry and accounting for enduring mistrust of domestic arms manufacturers among TNI leadership (Chairil, 2018), it is understandable that, while the TNI partners with domestic firms for a range of services (including maintenance and training) and lower-level procurement, it has historically preferred to purchase more advanced systems from foreign partners, including Russia and the Republic of Korea. In 2016, the (then) CEO of PT Pindad claimed that the industry is likely to need 5–10 years to reach the capacity to effectively produce globally competitive advanced platforms (Hermansyah, 2016). This is not unusual among ASEAN states, as none are among the top 25 arms exporting countries (Wezeman et al., 2018), despite being the second largest global weapons import market between 2007 and 2012 (Dowdy et al., 2014).

6.2.2.2 Foreign arms acquisition

Absent meaningful development and modernisation, the TNI will continue to rely on foreign arms suppliers to meet its demand for advanced weapon platforms, even if a given capability could be met by a domestic supplier. A recent example of this was the TNI's decision to purchase Italian helicopters in 2017 over a similar platform manufactured by PT Dirgantara Indonesia. Promoting the TNI's confidence in the reliability of domestic arms would be vital for any adoption strategy to succeed.

The TNI maintains a variety of significant arms purchasing relationships, which could be leveraged to purchase or gain access to autonomous military technology. Indonesia was the tenth largest global importer of arms between 2013 and 2017 (Wezeman et al., 2018), with its weaponry purchases rising sharply from US$36 million to nearly US$1.2 billion between 2005 and 2018 (Laksmana, 2018). Its main suppliers are the UK (17%), US (16%) and the Republic of Korea (12%) (Wezeman, 2018). Since 2014, the TNI has purchased three Chang Bogo-class attack submarines from the Republic of Korea and become a development partner on the KAI KF-X programme. In 2018 the TNI purchased 24 F-16C/D fighter aircraft from the US and has benefitted from US support under the Maritime Security Initiative to build its C4ISR capability and launch an unmanned aircraft squadron (Rahmat, 2018). Along with Russia, these states are developing autonomous military technology and have a track record of selling advanced weapon systems and transferring technology to the TNI.

However, it is also important to note that Indonesia's arms supply is also unusually diverse, with these top three suppliers only accounting for 45% of its total defence imports. Indeed, Indonesia ranked among the top three markets for five of the top 25 arms exporting countries between 2013 and 2017. For comparison, Vietnam (the 11th largest arms importer) relies on Russia to supply 82% of its arms (Wezeman et al., 2018). Recent major TNI purchases include a squadron of 11 Sukhoi SU-35 multi-role combat aircraft (Russia) and eight EMB-314 Super Tucanos aircraft (Brazil). This diversified supplier base is an interesting characteristic of the TNI's purchasing patterns and is arguably a holdover from the Cold War, during which the government was able to walk a fine line of neutrality to benefit from both major powers. There is also a clear concern among Indonesian policymakers that concentrating imports from a single supplier would impose vulnerability to embargoes, such as those previously imposed by the United Kingdom and United States over Indonesia's human rights record (Laksmana, 2014).

Despite the evident issues with Indonesia's ability to marshal its strong economic resources towards developing the resource capability to adopt autonomous military technology, there have been recent examples of the KKIP successfully coordinating the advancement of domestic arms production capability through the rigorous enforcement of technology transfer offset provisions in major procurement contracts and Indonesian military exports continue to grow, reaching US$284.1 million in exports between 2015 and 2018 (Grevatt, 2018). Returning to 2012, the Rheinmetall procurement deal was worth US$68 million more than PT

Pindad's total government income in 2013 (Silviana and Danubrata, 2015), a serious missed commercial opportunity. However, by 2018 PT Pindad had developed the Kaplan Modern Medium Weight Tank, a competitive medium tank designed in partnership with a Turkish firm and an Infantry Fighting Vehicle with superior capabilities to Rheinmetall's Marder IFV (Badak). Despite only completing its first successful live-fire test in August 2018, the TNI-AD has already expressed interest in replacing its stock of French AMX-1 light tanks with Kaplan Modern Medium Weight Tank, while Bangladesh and the Philippines expressed interest purchasing between 40 and 50 vehicles each (Nupus, 2018). In the maritime domain, PT PAL underwent a similar process with the SIGMA-class corvettes, originally purchased from the Netherlands (Bitzinger, 2013). A second contract in 2010 shifted production under license to PT PAL, who have subsequently been responsible for building the updated *Martadinata*-class frigate. If a process occurs with unmanned combat vehicles, even in partnership with another middle power, it is conceivable that the TNI would be able to eventually access a domestically produced (likely under license) autonomous weapon system. This contention is supported by the demonstration of an apparently supervised autonomous 'sentry gun' by smaller arms producer PT. Prafir Jaya Abadi and the Research and Development Agency in late 2018 (Pengembangan, 2018).

6.2.3 Organisational capital capacity

The second diffusion variable for consideration is whether the TNI possesses sufficient organisational capital capacity to adopt autonomous weapon systems. Horowitz describes three tests for measuring a state's organisational capital capacity, namely Critical Task Focus, Level of Investment in Experimentation and Organisational Age. The lower financial intensity requirement of AWS opens response options that have historically been unavailable to a smaller state like Indonesia; however, Indonesia's organisational capacity will still determine how the TNI will react to a LAWS demonstration point.

6.2.3.1 Critical task focus

It is apparent that the Indonesian civil leadership has a different Critical Task Focus (CTF) to that of the TNI (and particularly the TNI-AD). Developing a military power projection capacity that is reflective of Indonesia's perceived role as an emerging great power has been promoted as a priority by Indonesian policymakers. Contrastingly, the TNI focuses on improving its ability to respond to internal security threats and instability over regional power projection or external defence. While both positions would support the acquisition of some form of autonomous military technology, the TNI's CTF would require lower complexity platforms largely in the aerial and maritime domains, while larger warfighting platforms are required for the civil leadership supported CTF.

The focus on modernisation as a way to improve Indonesia's power projection capacity was reflected in the Minimum Essential Force (MEF) and Global

Maritime Fulcrum (GMF)[4] doctrines adopted by the Yudyono and Jokowai presidencies, respectively. The MEF was intended to be implemented in three stages: minimum essential force (2015–19), transitional essential force (2020–24) and ideal essential force (2025–29) (Haripin, 2016). MEF was intended to involve the rationalisation of TNI personnel numbers, an equal resource investment in the modernisation of weapon systems and platforms across all three branches of the TNI, and modernisation and development of the domestic defence industry, which included the purchase of 'Main Equipment Weapon Systems' (*Alutsistas*) (Haryanto, 2017). The latter GMF doctrine was originally touted during Jokowai's initial election campaign and had the broader focus of transforming Indonesia into a global maritime power (Arif and Kurniawan, 2018). The GMF was initially criticised for failing to provide specific guidance and operational direction to Indonesian security actors. While the long-awaited *Presidential Regulation (PERPRES) No. 16 of 2017 on Indonesian Sea Policy* was an improvement, issues remain with the operationalisation of the GMF into a workable strategic doctrine (Morris and Paoli, 2018). The fact that neither of these doctrines have been prioritised is indicative of a disconnect between the civil leadership and the TNI.

The TNI places a greater emphasis on ensuring the territorial and structural integrity of the Indonesian state than engaging in inter-state warfare. From a strategic doctrine perspective, the TNI acknowledges that there is comparatively little risk of a purely external invasion (Rabasa and Haseman, 2002). This is reflected in the Total People Defence System (Sistem Pertahanan Rakyat Semesta), which is based on a combination of guerrilla warfare, total warfare and outer island defence (Arif and Kurniawan, 2018), a sacrificial strategy intended to buy time to prepare a final defence of the island of Java. Rather than updating this policy, the TNI's 2015 Defence White Paper reaffirmed that internal, non-traditional and non-state threats (such as piracy, terrorism and political unrest) are still considered higher risk than inter-state conflict.

The split between the Indonesian state and the TNI's Critical Task Focus can be seen in the continued relegation of the *Tentara Nasional Indonesia-Angkatan Laut* (Navy, TNI-AL) and *Tentara Nasional Indonesia-Angkatan Udara* (Air Force, TNI-AU) to supporting roles, despite their centrality in both the MEF and GMF strategic concepts. The current operational doctrine for the TNI-AL (*Eka Sasana Jaya*) focuses on maintaining maritime security, supporting land-borne operations and improving regional military diplomacy (Agastia, 2017). Seemingly counter to the regional priorities of the GMF, the TNI has prioritised acquiring traditionally internally focused capabilities including frigates, corvettes and surveillance platforms. This reflected the status quo within the TNI-AL, where the capabilities of vessels as a proportion of the fleet remained largely constant between 2003 and 2015 and, with the notable exception of the three Chang Bogo-class submarines purchased from the ROK (Andersson, 2015), has not drastically shifted since. Furthermore, the TNI-AL has also had to divert resources away from modernisation and expansion to 274 vessels (under the MEF) towards maintaining some level of operational readiness in its current fleet. Far from preparing for pitched naval battles, the majority of Indonesian naval vessels play an active role in combatting a variety of non-traditional security threats and polic-ing territorial waters (Laksmana, 2018), a role that is reflected in the TNI-AL's strategic doctrine and, historically, its procurement pattern. This would indicate that the TNI-AL would be more successful in adopting smaller scale AWS, such as unmanned surface vehicles in a harbour defence role (Fleet Class USV) or unmanned maritime vehicles for long-term surveillance (Wave Glider), rather than more advanced LAWS. Given its current security risks, strategic doctrine and capability, it is clear that the TNI would gain less utility from the global strike capacity of a Taranis UCAV than from the tactical strike capability offered by AeroVironment's Switchblade loitering munition.

6.2.3.2 Level of investment in experimentation

While the majority of relevant research occurs in civilian universities, the Indonesian state maintains an important independent role beyond being the main source of research and development funding. Engagement between for-eign researchers and Indonesia is administered by the Ministry of Research, Technology and Higher Education, which has a purely administrative role and does not directly conduct research. Rather the relevant agency for military research is the Research and Development Agency (*Badan Penelitian dan Pengembangan*, also known as *Balitbang*), a division of the Ministry of Defence. *Balitbang* has four research subunits and an administrative Secretariat. One of these subunits, *Kapuslitbang Iptekhan Balitbang Kemhan*, is focused on evaluating, assessing and implementing research focused on emerging defence science and technology. *Balitbang* is also responsible for developing technical policies and is consulted on the administration of defence research funding. Aside from two research manu-scripts presented at an internal seminar in August 2018, there is limited publicly accessible evidence of ongoing research by *Balitbang* into autonomous military technology, with the majority of relevant research believed to be led by civilian researchers, typically in partnership with the national defence university.

Indonesia established its defence university, *Universitas Pertahanan*, in 2011 as part of the current modernisation effort. As with its foreign counterparts, the Indonesia Defense University is operated by the military and offers defence and strategic studies-related courses to TNI officers as well as the public. The Indonesia Defense University aims to become a world-class research institution by 2024 and has a track record of successfully partnering with other military universities. Reviewing the archives of its in-house journal, *Jurnal Pertahanan*, indicates a research focus on force modernisation, regional security and domes-tic security threats (such as separatism and terrorism). There were no references to autonomous weapon systems, although academics from this institution would play an important role both in developing an Indonesian operational praxis for AMT deployment and educating the next generation of TNI officers.

While publicly known engagement to date by civilian universities appears to be lacking compared to their counterparts in AMT developing states, the TNI and Ministry of Defence have been providing increasing support and funding to research students working on related technologies.[5] In 2017 the *Universitas Gadjah Mada* announced a research partnership with Indonesia Defence University. Although this was a broad partnership, the core focus was stated to be the development of defence technology including 'surveillance ship, sea robot, and rocket technology' (Marwati, 2017). Furthermore, *Bina Nusantara University* (Binus University) operates a Research Interest Group in Photonics and Computer Systems, which focuses on computer vision, photonics and computer engineering research. More recently, Binus University partnered with NVIDIA (best known as a manufacturer of computer graphics processing units) to open the American technology company's first artificial intelligence research and development centre. The centre will reportedly focus on developing deep learning technology for NVIDIA's Graphics Processing Units (Asian Scientist, 2017). While not as engaged as their foreign counterparts, Indonesian universities are evidently investing further in relevant research over time.

However, chronic underinvestment in military and civilian research and development has made it difficult to train, attract and retain the skilled personnel needed for military experimentation in this space (RSIS, 2014). While there are well-established, respected Indonesian research institutions, problematic privately operated higher education institutions comprise the majority of the higher education providers. More than 40% of university-level teachers were bachelor's degree qualified or less. Indeed, only three Indonesian universities were ranked among the top 500 globally in the QS World University Rankings in 2018 (Rosser, 2018). Without significant and long-term investment in research and development, the TNI will be unable to internally develop the personnel, capacities and structures needed to experiment effectively with emerging military technologies.

Despite commitments in the 2012 Defence Industry Law that 5% of future defence budgets would be devoted to research and development funding, modernisation and capital purchasing programmes continued to only receive a minority of defence spending. Military modernisation efforts have been stymied by chronic underinvestment within the TNI. This is partially the result of a disconnect between the funds allocated to the TNI by government budgets and the resources that it actually receives. This 'defence-commitment' gap is an enduring element of the Indonesian budgetary process (Sebastian and Gindarsah, 2011), occurring as recently as the 2017 revised defence budget, which was 11.8% lower than the proposal (International Institute for Strategic Studies, 2019). In 2011 there was an even more drastic gap of 58%, with the approved military budget being US$7.93 million lower than the proposed budget (RSIS, 2014). This gap is further exacerbated by a military bureaucracy that is riven by interdepartmental rivalry and struggling to improve transparency in the defence spending process (Sebastian and Gindarsah, 2013).

This limited actual resource allocation is further limited by the TNI's commitment to a force structure with unusually high operational, personnel and equipment maintenance costs. The TNI operates venerable equipment and platforms, obtained from a multitude of suppliers over the last three decades, leading to high operational costs and low readiness, averaging at between 30% and 80% in 2007.

Furthermore, the TNI remains personnel heavy; in 2010 the TNI spent approximately 44% of the annual defence budget on personnel costs and salaries, and an additional 22% on functional expenditure (Sebastian and Gindarsah, 2011). This meant that Indonesia allocated the second highest percentage to personnel costs (behind Mexico at 69%), while its operational cost expenditure was average among the MIKTA group (UNARM, 2018).

By 2018 the TNI allocated only 15.9% of the defence budget to research, development and procurement, which equates to roughly US$1.16 billion (International Institute for Strategic Studies, 2019). This actually represented a significant recovery in expenditure for the TNI's research funding, as between 2013 and 2018 its procurement allocation fell to only US$500 million (IHS Markit, 2018), while in the decade to 2014 IDR 8.32 trillion was set aside in funding for the domestic defence industry (Chairil, 2018). This recovery is expected to continue, with procurement funding predicted to expand to US$1.3 billion by 2022 (IHS Markit, 2018), as part of a total procurement and R&D investment of US$10 billion between 2018 and 2024 (Grevatt and Rahmat, 2018). Finally, a reduction in personnel expenditure to 50.2% of defence spending (the median among MIKTA members in 2009) could fund an increase in the procurement and modernisation allocation of roughly US$745 million, which could certainly fund more aggressive military experimentation efforts.

It is also worth noting that Indonesian businesses have emerged as meaningful contributors to its capacity to develop relevant technologies. For example, Kata .ai is developing the first algorithm for natural language processing of Bahasa Indonesia (Chitturu et al., 2017). In 2018 two reports were released that examined the level of engagement with artificial intelligence among Indonesian businesses. The first was a Forrester Opportunity Snapshot, commissioned by Appier (2018), which found that 65% of Indonesian business respondents had adopted artificial intelligence, while the second report (written by IDC) found a 24.6% adoption rate (Tao, 2018). The significant divergence in this data was likely due to the use of small sample sizes. Despite the statistical divergence, both reports agreed that Indonesian businesses were significantly further ahead in artificial intelligence adoption than their regional neighbours.

Overall, Indonesia's commitment to defence modernisation is evident from the steadily rising resource levels committed to its military. It would be defensible to assume that the allocation of these funds will be somewhat guided by the 'Minimum Essential Force' concept, which prioritises of modernising and strengthening capabilities rather than simply expanding manpower, given its centrality with current Indonesian defence planning (International Institute for Strategic Studies, 2018). However, aside from an admission in 2010 by the incumbent TNI commander of the need for a personnel 'rightsizing', the TNI remains reluctant to do so. Furthermore, comparing Indonesia's operational and personnel expenses with other ASEAN states demonstrates a similar, personnel-dominated expenditure pattern, which has been in place for more than 20 years (Laksmana, 2018). This path-dependency would limit the TNI's willingness to invest in experimentation towards increasingly autonomous weapon systems.

6.2.3.3 Organisational age

The final organisational capital variable identified in adoption-capacity theory is organisational age. Two variables are presented for evaluating this variable: the length of time since a military lost a major conflict or underwent regime change; and the nature of the domestic civil–military relationship (Horowitz, 2010). While the TNI has not been involved in any recent major interstate traditional conflicts, unpacking the civil–military relationship offers a valuable insight into the organisational age of the TNI.

While civilian governmental pressure is rarely sufficient to force militaries to innovate in a particular direction (Rosen, 1991), reviewing Indonesian military modernisation efforts since 2010 re-emphasises the presence of an unusual civil–military relationship. The TNI has historically been a major political actor, with its influence reaching a peak during the New Order period. The post-New Order period featured somewhat effective legislation and military reform that was designed to remove the TNI from direct involvement in politics. While the modern TNI is fiercely protective of its political neutrality in official documents, it also maintains its role in domestic policing, leading to (occasionally very public) conflicts with the national police (Almanar, 2017). Another contributing factor is the practice of 'territorial postings', which deploy TNI personnel alongside each level of government, even to the village level in local communities and internal security. As a result, the TNI remains a 'latent variable' in domestic Indonesian politics and has a complex relationship with the, theoretically superior, civilian Ministry of Defence (Laksmana, 2014).

This has translated into the military maintaining higher than expected influence over defence department decision-making. The higher echelons of the TNI remain largely strategically (and politically) conservative and committed to the Total People Defence System. Retired high-ranking military officers (primarily from the TNI-AD) (Arif and Kurniawan, 2018) have transitioned to senior roles in political parties, and even former President Yudhoyono was a retired General. Despite the introduction of the Global Maritime Fulcrum strategic concept, there remains no coherently outlined strategic plan for tri-service adoption of new advanced platforms. The unusually balanced civil–military relationship in Indonesia limits the capacity of the Indonesian government to pressure senior military officers towards the development or adoption of autonomous weapon systems.

6.2.4 Receptiveness of domestic audience

Unfortunately, there have been no publicly available studies conducted on Indonesian public opinion towards autonomous weapon systems at the time of writing, making it difficult to directly ascertain whether Indonesians would support LAWS. However, a 2014 Pew Research Center survey showed that 74% of Indonesians opposed the United States' use of remotely piloted UAVs, the direct precursor weapon system. In terms of indirect evidence, the Indonesian business community has a level of artificial intelligence adoption among the highest in the region, although this could lower public acceptance, given that approximately 56% of occupations in the Indonesian, Cambodian, the Philippines, Thailand and Vietnam economies are at risk of automation (Shewan, 2017). Finally, in November 2018, the Institute of International Studies at the Universitas Gadjah Mada became the first Indonesian organisation to become a member of the Campaign to Stop Killer Robots, signalling the first engagement of Indonesian civil society with the LAWS debate.

At a state level, Indonesia's engagement to date with the international discussion around potential regulation of LAWS is reflective of its broader foreign policy preference for neutrality. Due to Indonesia's refusal to become a signatory to the *Convention on Prohibitions or Restrictions on the Use of Certain Conventional Weapons which may be deemed to be Excessively Injurious or to have Indiscriminate Effects*, it has not been a direct participant in the ongoing international discussions at the United Nations. However, as a member of the Non-Aligned Movement (NAM), Indonesia's position can be inferred from NAM statements to the Group of Governmental Experts on Lethal Autonomous Weapon Systems in 2017 and 2018. The 2017 statement was delivered by Indonesian Ambassador Krisnamurthi, on behalf of the Non-Aligned Movement, to the General Assembly. However, using the descriptor 'statement' is perhaps generous; there was only a single relevant paragraph within the 15-page speech. Ambassador Krisnamurthi's speech covered a wide range of issues, including nuclear non-proliferation, biological weapons and the peaceful use of nuclear technology. In this paragraph, the Non-Aligned Movement, appears to support establishing an open-ended Group of Governmental Experts-led examination of the 'ethical, legal, moral and technical, as well as international peace and security related questions' raised by autonomous military technology, but stops short of supporting a pre-emptive developmental ban. This was followed by a more extensive statement in 2018, which was delivered by the Venezuelan delegation, and firmly established a preference for specific regulations and prohibitions being enshrined in international law in response to the emergence of LAWS, stating that any voluntary alternative 'cannot be a substitute for … a legally binding instrument'. This statement further called for a moratorium on development of LAWS pending the development of these international regulatory instruments. While not as strident as the positions taken by some other states, this statement clearly supported formal regulation, if not an outright developmental ban.

Given that the NAM consists of more than 120 member states, this could have been a major coup for supporters of a pre-emptive ban on LAWS; however, there are issues with relying on this statement to reflect the position of any individual NAM state. Firstly, in a similar manner to ASEAN, the NAM holds non-interference in the internal decision-making of member states as one of its founding principles. Secondly, the emergence of autonomous weapon systems appears to be low on the collective agenda of its member states. The 2018 NAM statement 're-emphasises' the position adopted at the XVII Summit of the NAM (2016) and 2018 NAM Ministerial Meeting (the Baku Declaration). However, the 206-page XVII Summit Final Declaration contains only a single paragraph that refers to LAWS, while the Baku Declaration does not actually mention autonomous weapon systems (Moncada, 2018).

Despite their shortcomings, these statements made Indonesia one of only three ASEAN member states to publicly state a position on the future of fully autonomous weapon systems. While these statements do not denote an independent Indonesian position or outline a definitive position beyond favour for further negotiation, Indonesia's prominent role in their formulation and delivery is reflective of its preference for soft-revisionism and strategic positioning.

6.2.5 *Capacity to Develop or Emulate Specialised Operational Praxis*

The final diffusion variable centres on whether the TNI has the capacity to develop or emulate a specialised operational praxis that will enable the full disruptive potential offered by autonomous military technology, up to and including Lethal Autonomous Weapon Systems. The operational praxis is effectively the process through which a given state military transforms a capability into force (Grissom, 2006). Examining the prior diffusion variables supports the conclusion that the TNI is unlikely to pursue fully autonomous weapon systems in a stand-alone combatant role, lacking both the resource capacity and the operational need to so. Instead, the TNI is likely to adopt an operation praxis that emphasises the resource efficiencies gained from deploying semi-autonomous platforms in internal security roles.

As outlined earlier in this chapter, the Indonesian military has not developed a sufficiently advanced national security innovation base to develop novel doctrine for the development and deployment of fully autonomous weapon systems. However, the government's commitment to mandating technology transfer in foreign arms purchases has stimulated the TNI to emulate and incorporate advanced weapon platforms acquired from other states. As an example, consider that five years after purchasing main battle tanks and infantry fighting vehicles from Rheinmetall, a domestic producer (PT Pindad) was able to offer competitive indigenous platforms to fill those combat roles. Furthermore, the Indonesian Ministry of Defence's Research and Development Agency has begun to promote its research efforts in this area, with a paper presented at an internal seminar in August 2018 and the unveiling of a prototype supervised autonomous weapon system in November 2018. While the TNI has not yet developed a specialised operational praxis internally, it is showing very early signs of interest and has demonstrated a capacity to emulate the praxes developed by other militaries as well as to effectively integrate technology transferred from foreign arms sources.

6.3 Evaluating Indonesian response options

As outlined in the methodology chapter, there are five response options available to states following the demonstration point of LAWS. First, the state could attempt to reassert its neutrality. Reasserting neutrality minimises the state's involvement in the incubation period and diffusion process but requires surrendering its agency, effectively 'hiding', until the international community or another actor resolves the shift in the current balance of power. This is arguably the most conservative and viable response option for smaller, middle and emerging power states in the Global South, at least during the current incubation period. Proclaiming neutrality, at least during the immediate post-demonstration point period, would reflect the strong penchant for hedging in both Indonesian and Singaporean foreign policy.

The second response option is to bandwagon with a first mover or early adopter, effectively attempting to gain protection or improved access to the innovation via association. The main benefits of joining with a great power state adopter would be greater procurement and technical support access, as well as a broader security guarantee. There are also the contingent economic and geopolitical benefits inherent in a great power alliance. Unfortunately, aligning too closely with a great power during a hegemonic competition could harm the middle power's relationship with the other hegemonic competitor (Fels, 2017).

The third and final external response option would be for the responding state to form a balancing alliance with other smaller states to 'offset' the advantage gained by the first mover (Horowitz, 2010). This would be a textbook defensive neorealist response for states that wish to retain their international influence, but individually lack the power and prestige to do so. A clear example of a balancing alliance was the formation of the Non-Aligned Movement to preserve the limited power of smaller states in the Global South during the Cold War (Fels, 2017). Initially adopting an external diplomatic response can be the most effective solution to protect the position of a state within the balance of power while its adoption capacity is insufficient to be an effective early adopter (Larsen, 2018).

A responding state also has two internal options to reassert its comparative position in the regional balance of power, it could attempt to catch-up by adopting the innovation (or an effective derivative) or it could attempt to develop a counter-innovation, a less resource-intensive advancement that offsets the advantage gained by early adopters (Larsen, 2018). Given the asymmetric nature of warfare and innovation, the latter arises almost organically, as the demonstration point forces rival states to aggressively pursue counter-innovations to offset their rival's initial first mover advantage (Gilli and Gilli, 2014). A recent example of a countering innovation was the 'carrier killer' missiles adopted by China (DF-21D) and Russia (Kh-47M2 Kinzhal) to offset the advantage the United States secured with its dominant carrier warfare advantage. In the case of autonomous weapon systems, effective offset strategies could include remote-operated unmanned vehicles, cyberwarfare, or purchasing and modifying Commercial Off the Shelf (COTS) platforms.

Attempting to become a secondary, fast-follower adopter is the final response option. Both Indonesia and Singapore have demonstrated a capacity to effectively emulate the operational praxes of great powers and to derive modernisation lessons from foreign manufactured advanced weapon systems that were integrated into their armouries. Furthermore, both states have acquired the capacity to domestically produce remote-operated UCVs, and successfully begun to integrate unmanned systems into their militaries, albeit to a greater extent in Singapore.

While following a primarily external response has traditionally been the only realistic option for middle power states, the lower initial entry barriers and rapid proliferation of remote-operated UCVs suggest that adoption, even on a limited scale, of increasingly autonomous weapon systems will be more feasible in this case.

6.3.1 Limited adoption

Although, as stated earlier in this chapter, Indonesia demonstrated an early prototype of a sentry gun with limited operational autonomous capability in late 2018, similar to those developed in the Republic of Korea (Parkins, 2015), the TNI is unlikely to become an early adopter of fully autonomous weapon systems. This is not unusual; as outlined earlier in this book, there are operational, political and technological barriers to any state developing or adopting a fully autonomous weapon system in the near future that would meet the definition used by this book.[6] As a smaller state with substantially lower resource capacity than, for example the United States, Indonesia is unlikely to independently integrate autonomous capability into its primary weapon platforms.

Based on its adoption capacity and demonstrated internal security focus, the TNI would be more likely to succeed by pursuing an operational praxis based in human–machine teaming. The TNI has the resource capability to adopt supervised autonomous weapons by purchasing 'off-the-shelf' platforms from a friendly state, although significant additional investment would be needed to support domestic production. For example, Indonesia has a history of purchasing fighter jets from Russia, most recently agreeing to purchase 11 SU-35s (Parameswaran, 2019). The clear upgrade from this platform would be the SU-57, an export version of which was promoted in late August 2019 (Udoshi, 2019). The SU-57 is a fifth-generation fighter that will reportedly be the preferred partner for the recently unveiled Russian S-70 Hunter-B, an armed UCAV designed for a similar operational role to the Boeing 'Loyal Wingman' programme. Even if the TNI-AU were not to purchase new aircraft to replace its ageing and disparate armoury, it could attempt to adopt semi-autonomous robotic pilots (like the South Korean Pi-Bot), artificially revitalising its air force. In a related example, the TNI-AU and the TNI-AL would both benefit from adopting 'swarming', low-cost autonomous vehicles for surveillance and intelligence operations.

Given the geographic nature of the Indonesian archipelago and the tenets of the Total People Defence System, teaming autonomous platforms with human supervisors to complete difficult, dirty or dangerous logistics and surveillance taskings would be both more strategically valuable and more likely to be successfully adopted. This approach would also allow the TNI to learn from cooperative exercises with foreign partners (like Australia and the United States). As the technology matures, it will also become more feasible for the TNI to purchase complete supervised autonomous platforms from the United States or Russia. Finally, the TNI's efforts to professionalise and modernise its command structures would benefit from adopting semi-autonomous battlefield assistants (emulating the British proposal), which would provide real-time analysis and advice for operational and tactical commanders. This would be particularly useful given the TNI's doctrinal emphasis on small unit delaying tactics across the archipelago in the event of an interstate conflict. Adding to the likelihood of successful adoption, each of these operational praxes could be achieved with technology (or even complete platforms) developed by other states, commercial entities or civilian researchers.

6.3.2 Counter-innovation/offset

Countering is particularly attractive to states, like Indonesia, that currently lack the capacity to adopt autonomous weapon systems but have the potential to do so over time, theoretically as part of the late majority (Larsen, 2018). Adopting a counter-innovation would also complement a response strategy that prioritised Indonesia's neutrality.

While Indonesia has the limited capacity to domestically produce UAVs, such as the Wulung (manufactured by BPPT Puna) (Donald, 2014), purchasing complete platforms would allow the TNI to bypass the majority of the research and development costs. Israel, the United States and China each produce remote-operated platforms that would suit the TNI's internally focused security doctrine. For example, the Chinese Caihong-4 has roughly comparable specifications to the MQ-9 Reaper with substantially lower barriers to purchase (Ewers et al., 2017). While from a maritime standpoint, Underwater Unmanned Vehicles (UUVs) have markedly lower operating costs than manned vessels; for example the Wave Glider UUV can remain deployed for up to a year, patrolling a section of ocean with visual and sonar sensors, for a fraction of the cost of a manned patrol boat (Mugg et al., 2016).

Indonesia would be able to further supplement its countering strategy by acquiring Commercial Off The Shelf (COTS) platforms from the myriad of state and civilian commercial providers. Given that civilian models have sufficient capability while being substantially less resource intensive than military platforms, their presence on the battlefield is understandable. For example, the current generation Mavic Pro, which is produced by the Chinese company DJI, has greater autonomous flight capability than the early model MQ-9 Reapers possessed and retails at roughly 1/5000th of the latter's price tag (Ewers et al., 2017). Due to their reliability, price and features, the Israeli military recently bought several hundred Mavic Pro drones for company-level formations to use for tactical ISR (Gross, 2017). While no serious comparison can be made with military models, Commercial Off the Shelf (COTS) drones are becoming ever more advanced, a factor that has already contributed to their use by violent non-state actors.

A more direct countering strategy would be for Indonesia to invest in improved cyber warfare capabilities, directly generating a capacity to undermine the deployment of more advanced autonomous and semi-autonomous weapon platforms. In recent years Indonesia has certainly improved its cybersecurity infrastructure, which was identified as a national priority in 2015 (Hanson et al., 2017). Indonesia operates two national-level Computer Emergency Response Teams (CERT), the Indonesia Security Incident Response Team of the Internet

Infrastructure/ Coordination Centre (ID-SIRTII/CC) (Setiawan, 2018) and the Indonesia Computer Emergency Response Team (ID-CERT). Despite this, there is no national cybersecurity policy and the TNI has only recently resolved to improve its cyber defence capabilities (Hanson et al., 2017). While the TNI should continue to develop into the cyber domain Indonesia would be better served by an offset strategy that prioritised the acquisition of remote-operated platforms from other states and civilian sellers.

6.3.3 Balance

Despite comparisons to the BRICS regional power states, it would be more effective for Indonesia to maintain a regional focus, collaborating with its fellow ASEAN member states to counterbalance the AMT developing states in the Asia-Pacific. There are meaningful benefits to be gained for Indonesia by combining its resource base and organisational capital capacity with other ASEAN states.

Although it is unlikely that even leading ASEAN member states will independently develop the level of C4ISR and data infrastructure (for transfer and storage) necessary to deploy unmanned platforms in the intercontinental manner of the United States, they certainly have the incentive and capacity to follow the example of China and Israel in deploying unmanned platforms with a less intensive internal security focus, which would reflect the TNI's critical task focus. This critical task focus is reflected in the fact that maritime assets are commonly used in non-warfighting roles, including law enforcement. Greater cooperation among ASEAN states becomes possible with unmanned platforms because they do not require the same level of operational security as manned platforms and militaries are willing to accept greater levels of risk with their deployment. Cooperative procurement, training and deployment would allow Indonesia and other ASEAN member states to increase the effectiveness of their efforts to combat the transnational organised criminal groups that operate in contested border waters.

Reasserting an alliance structure that is balanced against first mover states appears to be the most effective external response option for Indonesia to adopt. Cooperating with its fellow ASEAN states would allow Indonesia to grow its influence as a regional leader, protect its status in the shifting balance of power and continue modernisation efforts. However, cooperative development efforts and cross-training would increase the likelihood of semi-autonomous weapon systems proliferating to smaller ASEAN member states or even to regional non-state armed groups, necessitating a level of caution.

6.3.4 Bandwagon

Indonesia is unlikely to bandwagon with an early adopting great power due to the attendant risk of confrontation with the other hegemonic competitor and its long-term goal of limiting great power influence in the region. The relationship between the United States and Indonesia has shifted in importance and depth since Indonesian independence in 1949. Indeed, the United States was one of the countries that placed the TNI under a military embargo over human rights violations in the 1990s and early 2000s (Laksmana, 2014). US manufactured military hardware has therefore never prominently featured in the TNI's arsenal, representing only 10% of its platforms by 2017 (Wezeman et al., 2018). However, Indonesia recently secured a partnership with the US Navy to acquire advanced information and C4ISR systems, and features on the list of US partners exempt from sanctions for purchasing Russian weapons (Antey, 2018). While the Indonesian government has recently been making high-level defence diplomacy efforts, it has been careful not to offend China by supporting the United States-championed Freedom of Navigation Patrols in the South China Sea.

Like its fellow ASEAN member states, however, Indonesia maintains a lucrative economic relationship with China. In 2016 the rising superpower was Indonesia's largest export (11.62%) and import (22.71%) partner (The World Bank, 2018). While Indonesia has apparently not attempted to leverage this economic bond to procure autonomous military technology, China has demonstrated a greater willingness to sell armed unmanned combat vehicles, the precursor innovation to AWS, than the United States. As a result, Chinese systems have featured far more prominently in the market to date than their American counterparts, with armed variants already sold to Egypt, Jordan, Saudi Arabia, Iraq, Kazakhstan, Myanmar, Nigeria, Pakistan, Turkmenistan and the United Arab Emirates (Ewers et al., 2017).

Overall it seems apparent that this response option would not be in Indonesia's long-term interests because of the relationship cost that such an alliance would have on the other superpower. It would be much more effective to take a broader approach, strengthening existing arms and technology transfer agreements with the autonomous weapon developing states that currently do business with the TNI.

6.3.5 *Re-assert neutrality*

Re-asserting its neutrality would require that Indonesia take careful efforts to maintain balanced relations through the rising hegemonic tensions between the United States and China, who are both LAWS developing states with vital interests in Southeast Asia. Indonesia relatively successfully walked this line during the Cold War (with the tragic exception of the 1965–66 massacres of suspected Communists) (Bevins, 2017) and, as Ikenberry (2016) argues, smaller powers can reap significant benefits if they are able to balance the demands of two competing superpowers.

Adopting this approach would reflect the TNI's internal security focus and long-standing foreign policy goals. It would also enable the state to invest in matured autonomous military technology for relevant platforms in the early majority period, reducing the risk of an unsuccessful response. However, this approach would not be wholly reflective of President Widodo's push to modernise the TNI's capabilities as part of a wider push for Indonesia to be recognised as an emerging regional great power (Arif and Kurniawan, 2018).

While adopting Lethal Autonomous Weapon Systems would reflect Indonesia's nationalistic push to be recognised as an emerging regional power, the TNI's organisational capacity indicates that a limited adoption of semi-autonomous and supervised platforms is more likely to be successful and effective, although the impact on Indonesian prestige would be lower.

6.4 Conclusion

Notwithstanding Indonesia's economic strength, relative to other emerging states in its region, it is apparent that the TNI lacks the adoption capacity to effectively become an early adopter or first mover of fully autonomous LAWS. Despite high-level political support efforts to modernise the TNI, the domestic arms production industry and the national security innovation base have all been undermined by consistent underinvestment and do not reflect the critical task focus of the unusually powerful senior TNI-AD leadership. The Indonesian defence industry has, however, demonstrated a penchant for emulation which was supported by the implementation of a mandatory technology transfer provision in the Defence Industry Law 2012. Although it is unclear whether the Indonesian public would support the deployment of autonomous weapon systems, indirect evidence suggests that it is unlikely. Finally, the TNI has demonstrated an ability to emulate more advanced militaries and has been strengthening military education links to the United States, which is indicative of an ability to successfully emulate operational praxes for the deployment of autonomous military technology that were originally developed by a more advanced state.

Reviewing the array of response options open to Indonesia, it becomes apparent that the most likely response to successfully maintain Indonesia's comparative power and prestige in the shifting global balance of power is a combination of limited adoption, reassertion of neutrality and forming a balancing alliance, either within ASEAN or with other global south states. Based on its foreign policy preference for 'Pragmatic Equidistance' (Laksmana, 2017), reasserting neutrality (in a similar manner to the ongoing South China Sea disputes) would be an attractive and comparatively effective option, at least during the incubation period while modernising its power generation capabilities. Successfully adopting this combination of responses would give the TNI additional time to finalise its modernisation process and improve its doctrinal development. Given the nature of the TNI's organisational capital capacity, they are more likely to adopt less advanced platforms that have a lower adoption capacity threshold while retaining capabilities that would be effective in internal security roles, like surveillance, piracy interdiction and border security.

Notes

1　Mexico, the Republic of Korea, Turkey and Australia are the other members of the MIKTA partnership with similarly ranked economies.

2 *PT Dirgantara Indonesia* was named *PT Industri Pewsawat Terban Nesantara* (IPTN) until 2000.
3 These divisions are: planning, technology-transfer and offset, research and development, marketing and cooperation, finance, and legal.
4 This can also be translated into English as 'Global Maritime Axis'.
5 Personal author correspondence with a senior Indonesian defence academic.
6 'A fully autonomous Lethal Autonomous Weapon System (LAWS) is a weapon delivery platform that is able to independently analyse its environment and make an active decision whether to fire without human supervision or guidance' (Wyatt and Galliott, 2018).

7 Evaluating Singapore's adoption capacity

We must always fend for ourselves. No one will bail us out if we falter. In a rapidly changing world, this is one fact that will not change for Singapore.[1]

– Lee Hsien Loong (Prime Minister of Singapore)

7.1 Introduction

Among the ASEAN member states, Singapore would appear to have the most advanced capacity to adopt unmanned or autonomously operating military platforms. The Lion City is an important node in global commerce and a founding member of ASEAN. It maintains the most technologically advanced military among the ASEAN states and has a history of prizing military technology as a cornerstone of its deterrence-centred national security strategy.

The Singapore Armed Forces (SAF) is built around a core of highly trained regular soldiers, supplemented by conscripted reservists, supported by flexible and lethal air and naval forces. This structure is the culmination of the development of an advanced, albeit highly focused, domestic military industrial capacity, long-standing commitment to maintaining a technology-based military offset and the region's largest military budget. However, Singapore's lack of strategic depth, severe geographic restrictions, and demographic shifts have forced the SAF to develop the capacity to pre-emptively strike a threatening actor. These restrictions are encouraging Singaporean policymakers to consider the adoption of increasingly autonomous military technology.

Singapore complements the deterrent value of maintaining an advanced, compact hard force projection capacity, with the liberal application of soft power. Overall, Singapore maintains a strongly neutral stance between the two great powers that stems from a broader policy mindset that is remarkably defensive, exercising strict domestic control while proclaiming an 'unsentimental pragmatism' ideology that places the city-state above ethnicity, culture or language (Tan, 2015). Singapore's foreign policy approach can also be seen in the way Singapore leverages its participation and leadership within the ASEAN structure to maintain regional stability and the international rules-based order.

DOI: 10.4324/9781003172987-7

The purpose of this chapter is to critically evaluate whether Singapore has the potential to be a legitimate early adopter of LAWS. Given its advanced military, technologically superior civilian economy and respectably capable national security innovation base, it appears apparent that Singapore would be interested in adopting increasingly autonomous unmanned platforms. This chapter will carefully apply each of the five variables to demonstrate that Singapore possesses both a strong focus on increasingly autonomous systems and the capacity to successfully undertake a limited adoption of this innovation.

7.2 Evaluating Singapore's adoption capacity

7.2.1 Security threat environment

The physical defence of the city-state and its interests remains the main focus of Singapore's security services; however, the external focus of Singapore's economy and the growing threat of terrorism have added an increasingly non-traditional dimension to Singaporean security policy over the past two decades. This section will demonstrate how Singapore's threat environment will influence how the SAF perceives and would respond to the emergence of increasingly autonomous weapon systems.

The SAF's perception of its threat environment and the importance it places on maintaining a credible deterrent capability are clearly influenced by Singapore's early experiences as an independent state. Singapore's emergence in 1965 was characterised by multiple, potentially existential, threats. Gaining independence in the midst of the Indonesian–Malaysian Konfrontasi, Singapore inherited only a single infantry regiment, whose (predominantly Malaysian) officers could not be trusted. During the same period, Singapore's great power allies were otherwise engaged: the United Kingdom withdrew its presence east of the Suez Canal, which had been centred on its naval base at Singapore; while the United States was beginning to be pulled into what became the Vietnam War. Singapore's early leaders quickly identified that projecting a clear deterrent capability against potentially predatory neighbours was vital to Singapore's survival. The fledging SAF entered into, an initially secret, partnership with Israel, which provided not only the initial training and equipment, but also an early conceptual example of how to capitalise on a national service model to balance the need for a credibly sized military with the need to maintain economic growth within the constraints of a comparatively small population.

From a traditional perspective, Singapore's main security risk remains potential aggression or territorial infringement by a neighbouring state. Historically and geographically, this concern has focused on Indonesia and Malaysia. The SAF continues to hold that maintaining regional deterrence through a clear technological advantage is crucial for maintaining good relations with its closest neighbours (Yong, 2017). While relations and indeed defence cooperation between all three states have been improving since 2004 and 2003, respectively (Yong, 2017), the reignition of tensions around disputed territory near Tuas between Malaysia and Singapore in January 2019 (Rahmat, 2019) emphasised that fragility remains in these relationships. The adoption and integration of LAWS would somewhat off-set the population difference between Singapore and its neighbours, as well as reassert the SAF's military technology offset.

While defending against, and pre-emptively deterring, state aggression was clearly the first and most enduring goal of Singapore's national security policies, the current 'third-generation' SAF has prioritised smart, technologically assisted deployments of force to secure Singapore's regional interests. Arguably the most vulnerable of its vital interests are the long sea lines of communication (SLOC) through its territorial waters, through which vital trade transits (Yong, 2017). This leaves Singapore vulnerable to the disruption of these SLOC by a future state or non-state foe, which would do immense economic damage.

Singapore's economic reliance on uninterrupted, secure utilisation of the sea lines of communication in the region, even beyond its territorial waters, translates directly to a serious vulnerability to non-traditional security threats such as piracy and terrorism. Piracy offers a particularly problematic non-traditional security threat to Singapore because of its potential cost to recurrent trade and its inher-ently transnational and incorporeal nature. Despite the relative rarity of a serious incident, an attack on a merchant vessel can have major political and economic consequences. For example, the 2002 MV Limburg bombing resulted in a 300% increase in Yemeni ship insurance costs, which led to port volumes being slashed by 50% (Farley and Gortzak, 2009).

Beyond the monetary cost of its predation on maritime shipping, modern piracy shares very little in common with its historical predecessor. In a definitional sense, modern 'piracy' can be roughly equated with maritime robbery. The majority of 'piracy' incidents could be more accurately characterised as petty theft and occur while the ship is docked in port. When incidents occur at sea, they are generally opportunistically carried out by small groups of poverty-stricken fishermen with threats of violence being far more common than inflicted violence (except in the case of small target vessels). Notably rarer, high-value pirate attacks are generally well-funded and well-planned,[2] reflecting the participation of sophisticated crimi-nal syndicates and corrupt local officials (Hoesslin, 2016). In comparatively few cases, the crew of ships taken by pirates are sometimes killed, their vessel later reappearing under a new identification for use in black market trading.

The Singapore Armed Forces' response to piracy has incorporated a greater emphasis both on multinational and regional cooperation and 'hardening' its phys-ical security. The Singapore-led *Regional Cooperation Agreement on Combating Piracy and Armed Robbery against Ships in Asia* (ReCAAP) was one of the more successful efforts to bridge the interstate tensions and barriers that had under-mined prior efforts. ReCAAP introduced a multilateral information sharing and coordination service that is centred on a 'fusion centre', which is operated by the Singaporean Navy (Haacke, 2009). Incorporating increasingly autonomous mili-tary technology offers more resource-efficient platforms for harbour defence and long-term maritime surveillance, which are valuable tools for deterring piracy.

The Singaporean Navy became one of the first states to deploy an unmanned surface vehicle in 2005, adopting the Protector USV, designed by Rafael Advanced Defense Systems, which is remote-operated (with limited autonomy), highly manoeuvrable and capable of being armed. While its primary purposes are surveillance and force protection (in harbour), the 2017 third-generation model demonstrated the capacity to fire Spike ER missiles (Williams, 2017).

For the Singapore Armed Forces and the Ministry of Home Affairs (its civilian counterpart security service), responding to the high risk of terrorism is a major strategic and operational priority. A secular democracy with close western ties, Singapore has been previously named as a target by Al Qaeda and ISIS, as well as regional affiliates. Singapore is not a stranger to terrorist attacks, suffering at the hands of Indonesian saboteurs during the *Konfrontasi*; however, Singapore has not experienced a successful terrorist attack in the post 9/11 era. Equally though, the risk of Islamic terrorism has evolved during this time. Singaporean security forces were originally most concerned by Al Qaeda and local affiliated groups (such as Jemaah Islamiyah [JI] and the Abu Sufyan group), which primarily focused on western-affiliated targets, such as embassies and nightclubs that were popular with foreigners. The rise of ISIS shifted the threat to local groups, which were inspired and radicalised by the local affiliate and planned transnational attacks. By 2015, approximately 19 organisations in southeast Asia were suspected to have pledged loyalty to ISIS (Gunaratna, 2017).

The response by Singaporean security services was a combination of regional diplomacy and intelligence sharing, surveillance, cooperation with the local Muslim community and the judicial application of preventative detention and control orders. Between 2015 and 2017, 14 Singaporeans were placed under preventative detention under the Internal Security Act, while 40 Bangladeshi nationals and 8 Indonesians were identified as having been radicalised and subsequently deported (Ministry of Home Affairs, 2017). While the prospect of militants or radicalised individuals coming to Singapore remains a concern, the Ministry of Home Affairs has identified returning foreign fighters and radicalised homegrown lone wolves to be the main risk (Ministry of Home Affairs, 2019). To this end, Singaporean authorities have made strong but ultimately unsuccessful efforts to progressively shut down internet sources of radicalisation and training.

While at first glance increasingly autonomous military technology may not seem to have a place in addressing the risk of terrorism, however, on closer examination AMT offers clear counter-terrorism value for Singapore. Unmanned platforms are notably more resource-efficient for active surveillance and can be deployed without human risk into dangerous or difficult-to-navigate sections of coastal waters to interdict the entry of militants and equipment. More importantly though, autonomous technology offers a solution to the tyranny of data problem, which was also encountered by the British intelligence agency GHQ. A key barrier to effective mass surveillance of potential radicalised citizens is the sheer number of man-hours required to review and analyse gathered intelligence and data. Artificial intelligence research offers the potential for autonomous systems to process raw meta-data at a far higher and more reliable rate than humans.

Utilising pattern and facial recognition, guided by machine learning, an artificial 'assistant' would be able to flag potentially suspicious behaviour for a human analyst. Finally, in the event of a terror attack, autonomous weapon platforms offer the ability to deliver accurate firepower, dispose of explosives or rescue casualties in situations that are too dangerous for human first responders. The United States and ROK are both developing AWS that could be utilised in these roles, and US domestic law enforcement have already utilised remote platforms in armed offender incidents.

In contrast to larger state developers of LAWS, Singapore's threat environment places a far greater emphasis on non-traditional security risks, particularly the economic and political cost that would result from a major terrorist attack. Even a purely traditional security view of Singapore's security environment necessitates a posture that prioritises 'smart power' and a forward deterrence capability, offsetting the SAF's smaller size and Singapore's lack of strategic depth with the demonstrated capability to pre-emptively strike and degrade an aggressor in their own territory. Therefore, Singapore's security environment would not require, or necessitate, that the SAF emulate the United States in its pursuit of LAWS with a global strike capability. Of more value would be smaller-scale platforms and systems that improve the individual lethality and survivability of SAF units or bolster the capacity of Singaporean security services to anticipate, surveil and thwart home-grown terrorist threats.

7.2.2 Resource capacity

The Singapore Armed Forces are generally considered to be the best equipped among Southeast Asian militaries, which reflects strong, consistent and carefully targeted defence spending. Arguably to a greater extent than Indonesia, Singapore is well placed to adopt increasingly autonomous weapon systems from a purely economic standpoint. Both states have enjoyed strong economic growth, invested in military modernisation and maintain valuable foreign arms-transfer partnerships. However, the two states diverge at this point, making Singapore's resource capacity significantly higher than Indonesia's. Singapore has an innovative, advanced and respected domestic military industrial base, and its procurement decisions are guided by a well-resourced national security innovation base. Incorporating increasingly autonomous military technology offers a clear and appealing solution to the increased pressure from an ageing, declining population on the SAF, which is largely comprised of conscripts, stiffened by veteran officers and advanced technology.

Singapore is a major economic centre whose wealth has historically been largely dependent upon international trade and commerce. While the Singaporean GDP is 36th highest globally, from a per capita perspective its ranking rises to 7th (CIA World Factbook, 2019). A discrepancy that highlights that Singapore is beginning to feel the effect of ageing on its declining population, which is already small by regional standards (Jamrisko and Amin, 2017). Although structural growth slowed over the past five years (Forbes, 2018), the Singaporean economy is still steadily expanding, maintaining a year-on-year average growth rate of 3.4% in 2018. Driven by innovation and technological improvement, this growth was predicted to continue in 2019 despite global trade tensions (Economic Policy Group, 2018). These tensions present an economic risk for the export-oriented economy, which is Southeast Asia's most influential technology and finance hub, a status that is reflected in Singapore's decision to join the Comprehensive and Progressive Agreement for Trans-Pacific Partnership and Regional Comprehensive Economic Partnership.

Reflecting the importance it places on military deterrence, Singapore has consistently maintained an unusually high defence budget relative to its population and economy size. Singapore accounts for 2.7% of defence spending in Asia, a figure that poorly reflects the fact that Singapore accounts for significantly more arms purchases than its ASEAN peers, the closest competitor being Indonesia at 1.8% (International Institute for Strategic Studies, 2019). Indeed, Singapore had the highest military expenditure in Southeast Asia in 2017, with a total defence budget of SGD 14.2 billion (USD 10.2 billion) (International Institute for Strategic Studies, 2018). This was a comparably minor increase in real terms over the 2016 allocation and reflected a 0.1% reduction as a percentage of GDP (Defence Intelligence Organisation, 2018). However, the SAF received a 3.9% funding increase on 2017 levels in the 2018 budget, with defence spending rising to SGD 14.76 billion (USD11.2 billion) (International Institute for Strategic Studies, 2019). Although military spending as a percentage of GDP has declined steadily from 4.8% in 2005 (Matthews and Yan, 2007) and 4% in 2008 to 3.2% in 2017, this is still higher than the international average and equates to roughly 19.8% of government spending (Defence Intelligence Organisation, 2018). The electoral dominance of the People's Action Party (PAP) in domestic government has guaranteed this level of funding over the long term (Laksmana, 2017), with defence spending growing consistently by a total of USD $1,231 million since 2008 (in constant 2016 USD terms).

However, it will become increasingly difficult for Singapore to maintain the distinction of having the highest military expenditure in the region over time simply because its economy is not growing at a comparable rate to its closest competitors. The Singaporean economy is already ranked fourth among ASEAN states and the gap between Singapore's military expenditure and the rest of ASEAN has been narrowing since 2006. Although the government committed in 2018 to maintaining defence spending at 3–4% of GDP in line with inflation (Zhang, 2018), this is substantially lower than pre-2002 levels which ranged from 3.7% to 5.68%. A technological offset strategy is substantially less effective when potential rivals have higher military purchasing power, which they can capitalise on to acquire comparable capabilities or a counter-innovation.

As an illustrative example, Indonesia would only have to raise its defence expenditure, currently at USD 8.1 billion (0.8% of GDP and 5% of government spending) by 21% to surpass Singapore. Although this sounds like a major increase, Indonesia's military expenditure rose by 122% between 2008 and 2017 (Tian et al., 2017), driven by a significantly higher average economic growth rate

(Defence Intelligence Organisation, 2018). On the other side, Malaysia's military expenditure doubled between 2000 and 2013 (Bitzinger, 2015), although it was then cut by 13.7% in 2016 (Defence Intelligence Organisation, 2018). While Singapore has historically been able to leverage its greater military spending to maintain the technological offset at the core of its strategic outlook, the comparatively greater economic potential of its neighbours means that this is becoming less viable (Tjin-Kai, 2012).

7.2.2.1 Domestic military industrial base

Singapore's domestic military industrial base plays a crucial role in maintaining the SAF's technological advantage over its neighbours (Bitzinger, 2018). Its form is a result of a combination of consistent military spending, a strongly hierarchical and controlled society, and multiple linkages to an advanced, innovative civilian economy. Singapore's military industry is consistently referred to as one of most advanced in the region by organisations such as McKinsey and Company. However, capacity is distinct from active capability and Singapore's defence industry is currently focused on niche production, supplemented by an openness to foreign investment and commitment to evolutionary platform improvement that is unusual in the region. While Singapore possesses the capacity to produce autonomous military technology, potentially even weapon platforms, this is not reflected in its current arms production.

Singapore's domestic arms industry is dominated by the state-owned Singapore Technologies Engineering (STE). STE was the 57th largest arms exporting company globally in 2017, a drop of five places from the year before (Fleurant et al., 2018). This translated to USD 1,680 million in total arms sales. However, STE is a highly diversified company, with over 200 partially state-owned subsidiary entities (Matthews and Yan, 2007), and arms sales accounted for only 35% of STE's total sales in 2017. STE has invested significantly in Irish and American defence-related manufacturers, and almost a quarter of its workforce is based outside of Singapore (Fleurant et al., 2018). Singapore is unusual among ASEAN states in that it does not typically require that a technology transfer component be included in foreign arms partnerships, and (unlike Indonesia) has been actively encouraging foreign investment in military-related development efforts (Tan, 2013).

Singapore Technologies Engineering has four major component companies that are worth noting, each services a niche market within one of the three military domains. The first, ST Aerospace is interesting in that it does not build aircraft; rather it is responsible for maintaining and upgrading the SAF air fleet. However, ST Aerospace was also responsible for developing the Skyblade IV, a tactical surveillance UAV. ST Aerospace also conducts commercial maintenance, which accounted for a significant portion of ST Aerospace's S$ 2.06 billion income from contracted work during 2018. ST Marine is responsible for building and maintaining the Republic of Singaporean Navy's vessels, and has developed a notable track record of winning commercial and military export contracts. Recent ST Marine contracts include producing offshore support vessels for a commercial maritime oil extractor in the United States and patrol vessels for Oman. ST Marine has also demonstrated a capacity to learn from the designs of foreign shipbuilders to inform their design of new vessels for the RSN or upgrades for existing warships. The land division of STE, ST Kinetic, develops tracked combat vehicles and armoured personnel carriers, as well as ammunition and artillery. Aside from supplying the SAF, ST Kinetic has enjoyed limited success securing export controls with several foreign militaries including the United Kingdom. The final component firm is ST Electronics, which focuses on electronic warfare, signals intelligence and communications (Tan, 2013), which would be of direct relevance to developing autonomous weapon systems. ST Electronics has assumed a commanding position in the market for Very Small Aperture Terminal satellite components and plays an important role in developing the SAF's Command, Control, Communications, Computer, Intelligence, Surveillance and Reconnaissance (C4ISR) capabilities, a vital component for a state to sustain a large-scale deployment of unmanned platforms. Crucially, each of these subsidiaries has established a place for itself in the civilian market for versions of their primary product, such as construction equipment (Tan, 2013).

The SAF takes a long-term view when determining its arms procurement priorities, which affects the production and development of arms and military-related technology, the industry's investment decisions reflecting the SAF's identified goals. The Defence Science and Technology Agency, a corporatized government entity, plays a major advisory role in determining these priorities as well as in military research and development. Acting as a kind of technocratic gatekeeper, the DSTA identifies emerging military technologies that would be of interest to the SAF and determine the feasibility of developing the needed capability domestically. It is worth noting that the SAF has demonstrated a capacity to purchase commercial off-the-shelf platforms, where it is either more efficient or diplomatically expedient than procuring from a domestic supplier.

In what is arguably a sensible economic decision, Singapore's arms exporters have generally avoided putting themselves in direct competition with major foreign arms manufacturers. While the city-state would theoretically have the technological capacity to become a minor but active state developer of autonomous weapon systems (similarly to South Korea), it has contented itself with aiming for niche markets, such as ammunition, small arms and light armoured vehicles (Bitzinger, 2017). Singaporean arms have been sold within Southeast Asia as well as more globally, including to the United Kingdom, Nigeria and Brazil.

However, arms exports are only supplementary to the core purpose of domestic arms production, which is its contribution to 'Total Defence', ensuring the SAF's technological advantage and targeting its production capacity towards emerging niche military export markets. Singaporean firms supply most of the lower-level equipment utilised by the SAF, and it is one of the few domestic arms industries in the region that have the capacity to produce more complex platforms like medium howitzers and battleships (Heiduk, 2017).

The SAF and defence industry aim to undergo steady, evolutionary innovation, improving existing and emerging platforms with a long-term investment plan.

In this vein, Singaporean arms manufacturers perform the majority of required maintenance on procured platforms (Bitzinger, 2017) and have developed a reputation for very successfully upgrading or retrofitting these platforms. For example, Singaporean firms upgraded the weapon delivery and navigation capabilities of Northrop Grumman F-5E Tiger II aircraft (Matthews and Yan, 2007). These improvements are seen as a crucial component of Singapore's ability to maintain the 'technological edge' necessary to offset their size deficit.

7.2.2.2 Foreign arms acquisition

Befitting its investment in the international economy, Singapore has a well-established track record of procuring advanced platforms from foreign partners in response to identified capability gaps. Indeed, the first generation of the SAF was shaped to a large degree by the transfer of military knowledge and equipment from Israel. The modern SAF relies instead on the United States for the majority of its foreign arms acquisitions. This is unsurprising given the clear influence that the United States' adoption of information-warfare had on the SAF's '3G fighting force', the conceptual umbrella for its modernisation efforts (Laksmana, 2017).

While Singapore's importance as a weapons import market had declined relative to its regional neighbours by 2017, it nevertheless remains a significant importer. Singapore had invested significantly more resources and accounted for more foreign arms purchases between 1988 and 2009 than any of the other ASEAN member states. Despite dropping to the 21st largest importer of arms position by 2017, Singapore still accounted for 1.5% of global arms sales between 2013 and 2017.

Unlike its Indonesian counterpart, the SAF purchases the majority of its arms from the United States, which accounted for 70% of purchased platforms during this period. For example, the SAF purchased 16 F-15SG combat aircraft from the US and expressed interest in joining the Joint Strike Fighter program. The remaining 30% of arms were more evenly sourced, with France and Italy accounting for 12% and 4.1%, respectively (Wezeman et al., 2018). Befitting an island state, these procurements were largely destined for the Singaporean Navy and Air Force, a pattern that remained stable throughout the 1990s and 2000s (Tjun-Kai, 2012). For example, the SAF purchased 120 French MICA missiles to be installed on a class of eight corvettes, which were to be built locally but included components sourced from European defence firms (Heiduk, 2017).

While Israel is no longer the main source of foreign arms for the SAF, the Lion City still maintains a relationship with Israeli defence firms, which is particularly relevant to this analysis because it is from Israel that Singapore purchased its current inventory of remote-operated unmanned vehicles. As discussed earlier in this book, Singapore has partnered with Israel to complement its domestic production of remote-operated unmanned vehicles, purchasing Heron and Hermes MALE UAVs, as well as the lighter Searcher MkII. This cooperation extends to the maritime domain with the Protector USV, originally designed by Rafael Advanced Defense Systems, which was initially deployed by the Singaporean

Navy in 2005 and, like most platforms adopted by the SAF, has been subsequently upgraded. While Singapore has not expressed direct interest in procuring armed UAVs from Israel or other providers, this is likely to avoid antagonizing or threatening their neighbours rather than a lack of capacity (Desker and Bitzinger, 2016).

The historical pattern of foreign arms procurement by the SAF suggests that it would be feasible, and not particularly remarkable, for the SAF to purchase increasingly autonomous platforms for deployment in the maritime and aerial domain. However, in the absence of widespread (or threatening) adoption of armed autonomous platforms by its regional rivals, Singapore is unlikely to violate its doctrine of 'strategic restraint' by purchasing lethal or offensively oriented autonomous weapon systems (Heiduk, 2017). Rather its existing foreign arms relationships and procurement patterns would support the procurement of semi-autonomous or supervised autonomous platforms in the maritime and aerial domains, which can be used in a deterrent role as well as for improving maritime security. This would also support Singapore's current shift away from a deterrent strategy to a more diplomatic approach that prioritises regional cooperation on transnational security issues such as terrorism and piracy (Tan, 2015). That autonomous platforms offer this capability at a lower resource cost than advanced manned platforms will be increasingly important given the difficulty of maintaining a technological offset against a neighbouring state that will soon be able to invest greater financial resources into securing advanced weapon imports.

7.2.3   Organisational capital capacity

The second diffusion variable for consideration is whether the SAF possesses sufficient organisational capital capacity to adopt autonomous weapon systems. Horowitz describes three tests for measuring a state's organisational capital capacity: critical task focus, level of investment in experimentation, and organisational age. The lower resource capacity required to adopt AWS opens response options that have historically been unavailable to smaller states; however, Singapore's organisational capacity will still influence how the SAF will react to a LAWS demonstration point.

7.2.3.1   Critical task focus

The critical task focus of the Singapore Armed Force has undergone three major shifts since the city-state gained independence, yet the importance of identifying and adopting emerging military technology in offsetting Singapore's lack of population and strategic depth has consistently retained a prominent position in state and military position statements and doctrine. Unlike its Indonesian counterpart, the SAF and civilian government hold similar views on the importance of emerging military technology to its core functions. While Horowitz originally argued that a strict, well-defined critical task focus can limit innovation, the opposite appears to hold in the case of the SAF, whose planners have already drawn a clear link between the fourth-generation SAF strategic concept and increasingly autonomous weapon systems.

While robotics and unmanned platforms only began to feature in discussions of the third-generation SAF (3G SAF), technology has always been a central element of offsetting the SAF's structural disadvantages. The first-generation SAF was focused on rapidly building a deterrent capability, relying on advice from the Israeli army (although the extent to which this relationship was 'reliant' has been challenged by Singaporean officials in recent years). The first-generation SAF relied on advanced platforms adopted from foreign suppliers to stiffen its core of inexperienced national servicemen. This was necessary for the SAF to present a credible deterrent threat towards its neighbours, itself the basis of the 'poisoned shrimp' strategic outlook. This view of Singapore's defence recognised that Singapore was vulnerable to attack and instead aimed to ensure that the city-state would be seriously difficult to 'digest', promising to embroil an aggressor in ruinous urban guerrilla warfare (Tan, 2015). This was an image that was actively promoted by Singapore's leaders and unfortunately inspired the moniker of the 'Israel of Southeast Asia'. While not completely accurate even during the 1960s (Tan, 2015), it did reflect that the Lion City was willing to put the state's survival above all other concerns.

The second-generational shift in the SAF occurred in the 1980s, when Singapore was more firmly established and had begun to grow economically. Reflecting the city-state's shifting security environment and resources, the second-generation SAF adopted a more conventional force structure, upgraded and expanded its stock of advanced foreign weapons and established the early elements of Singapore's national security innovation base. This evolution was accompanied by a shift in strategic outlook, with the adoption of the 'porcupine' strategy. While retaining, even emphasising, technological offset-based deterrence, the 'porcupine' strategy introduced a pre-emptive element, envisioning the capacity for the SAF to pre-emptively strike an aggressor within their own territory, reducing the challenge posed by Singapore's complete lack of strategic depth (Raska, 2015). This involved assigning a greater role for the Republic of Singapore Air Force (RSAF) and Republic of Singapore Navy (RSN), who now assumed a limited power projection role within Singapore's maritime territory and nearby waters (Raska, 2015). This increased role was reflected in the significant resource investment into procuring and upgrading naval and aerial platforms during the 2000s; for example, Singapore was the only ASEAN member state included as a partner in the Joint Strike Fighter programme (Bitzinger, 2010).

The modern, third-generation Singapore Armed Forces reflects the economic power of the Lion City at the height of its importance as a hub of global commerce, while also reflecting the challenges of violent non-state actors and the closing gap between its defence spending and that of other Southeast Asian states, especially Indonesia. The third-generation modernisation was challenged by the difficulty of justifying high military spending against the demands of the civilian economy. Although the ruling PAP has historically been able to legitimise consistently high defence spending, this has been challenged in recent years. As a result of these challenges, the SAF has adopted the 'dolphin' strategic outlook as part of its post-transition continuing modernisation, envisioning a smart, technologically enabled and operationally networked military with a greater power projection capability (Raska, 2015). This is balanced by a renewed emphasis on 'soft power', securing Singapore's interests through defence diplomacy, regional security arrangements and committing to a greater role in multilateral organisations (Tan, 2015). Reflecting the 'top-down' nature of Singaporean military innovation to date, the development of 3G capability was meticulously set out as a three-stage process that reflected the importance of combining the adoption of emerging military technology with the development of operational concepts for its effective integration.[3]

The emergence of LAWS will test whether the SAF has sufficiently improved its capacity to adopt disruptive innovations. There is certainly evidence that the SAF is interested in developing and adopting increasingly autonomous military technology. The Ministry of Defence established the Future Systems Directorate with the explicit goal being to 'push the boundaries' of operational concepts and new military technology, a mission that carried over following the merger into the Future Systems and Technology Directorate (Raska, 2015). By the mid-2000s, the SAF had invested substantial resources into developing operational concepts and integrating niche advanced weapons platforms, including unmanned platforms like the Protector USV. More generally, unmanned platforms have become linked to the 3G SAF, particularly unmanned aircraft, because they offer a resource-effective capability that can be quickly adapted for use in a defensive conflict (Desker and Bitzinger, 2016). Furthermore, autonomous systems were linked to the emerging fourth-generation SAF (Next-Gen SAF) by Singapore's defence minister as early as 2015 (Raska, 2015). Finally, given the centrality of maintaining a defensive offset against Indonesia, it is worth noting that attempted adoption of AWS by either state is expected to trigger a reflexive attempt by the other.

In contrast to Indonesia, the adoption of autonomous military technology offers capabilities that reflect the critical task focus of the SAF, that is, leveraging emerging technology to offset the resource and strategic depth constraints faced by the SAF. However, given the characteristics of the dolphin strategy, it is unlikely that Singapore would pursue armed or otherwise 'offensive' AWS in the absence of provocation.

7.2.3.2 Level of investment in experimentation

Both domestic and foreign arms acquisitions by the SAF are guided by the comparatively advanced national security innovation base. From its inception the SAF has incorporated high-level civilian oversight and contribution; indeed, the early SAF could have been described as 'civil-servants in uniform' (Laksmana, 2017). This established a track record of cooperation between military and civilian personnel that was reflected in Singapore's efforts to maintain its technological edge. Indeed, Singapore's national security innovation base, which is generally referred to domestically as the 'defence technological community', is regularly described as the SAF's 'fourth service' branch.[4]

The importance of maintaining the SAF's technological offset is reflected in the consistent funding allocated to this 'fourth service branch'. In a similar practical concession to that made by its domestic arms industry, the Singaporean government has recognised that it is impossible to stay ahead of its rivals in all relevant technologies, and it therefore carefully directs its research targets at niche areas that complement civilian research, while utilising development partnerships with allies to address important identified shortfalls (Goldman and Mahnken, 2004).

Unfortunately, specific statistics on the division of Singapore's defence budget in more recent years, allocations of subsidiaries to civilian arms producers, and the details of current arms deals are generally not publicly available (Tan, 2013). However, in the absence of official government data, this analysis can draw upon estimates and extrapolated data from published scholarly literature and corporate research documents to demonstrate that the SAF has consistently allocated significant resources to funding military research, development and procurement.

Reflecting its preference for 'spiral' evolutionary innovation, SAF military spending has gone through several cycles. The most recent significant uptick occurred during the transition to its third generation with funding levels reaching up to 6% of GDP (Mapp, 2014). During the same period, defence research and development spending rose from 1% to 4% percent, which equated to an increase of USD 140 million (Goldman and Mahnken, 2004). Consistent spending during this period ensured that the current SAF's equipment is more modern than their counterparts in a region that was characterised by aircraft stocks purchased in the 1970s–80s, and a significantly higher percentage of the SAF's platforms are considered to be operation-ready than the regional average. The latter also reflects the fact that the SAF has a demonstrated preference for upgrading existing platforms instead of purchasing replacements (Wyatt and Galliott, 2018).

In terms of unpacking Singaporean military expenditure, the 2018 budget allocated USD 2.7 billion (approximately 25%) to the maintenance and upgrading of existing platforms and USD 5.3 billion (approximately 49.12%) to personnel costs. Both allocations are expected to increase minimally as a percentage of military spending (0.2% in the case of the personnel allocation and 0.4% for maintenance) by 2022. Outside of generational transitions, the SAF has generally maintained a stable allocation of funding to research, development and procurement. In 2018, research and development accounted for 4% of total defence spending, which translated to USD 340 million per year (Jane's Sentinel Security Assessment, 2018), while procurement is estimated to account for 13–16%, contributing to a combined allocation of USD 2.18 billion (International Institute for Strategic Studies, 2019). Within this expenditure, the Republic of Singapore Air Force is expected to account for 34% of total Southeast Asian spending on air force modernisation between 2018 and 2022, with a predicted outlay of approximately USD 4.2 billion (Jane's Aerospace, Defense and Security, 2018). Singapore is currently regarded as the most prolific funder of research, development and procurement in Southeast Asia, which further supports the contention that the SAF would be well placed to become an early secondary adopter of autonomous military technology.

Beyond simple financial investment, the importance of experimentation and research to the SAF is further reflected in institutional terms by the fact that three active major military research organisations support its innovation and procurement. These agencies broadly focus on the research and development of defence technology (Defence Science Organisation National Laboratories); developing innovative operational concepts (Future Systems and Technology Directorate); and coordinating the innovation, development and procurement process for the SAF (Defence Science and Technology Agency). In 2017 Singapore announced an additional annual investment of USD 32 million to create new research laboratories as part of an effort to promote autonomous system, data analysis and artificial intelligence research within the defence technological community (Grevatt, 2018).

The oldest member of this community is the Defence Science Organisation, which was only publicly acknowledged in 1989, 17 years after its establishment. Renamed to the DSO National Laboratories during a corporatisation process in 1997, the DSO is currently the largest defence R&D agency in Singapore with over 1,500 engineers and scientists spread across ten organisational divisions.[5] Particularly relevant divisions to this analysis are 'Emerging Systems', 'Electronic Systems' and 'Guided Systems', although others focus on sensor technology and information systems. In terms of specific research laboratories, it is worth noting that the DSO operates the UAV System Integration Reliability Laboratory, which is utilised by both military and civilian researchers in their development of advanced unmanned aircraft. In addition, in mid-2017, the Singaporean defence minister announced additional funding for the DSO to open a robotics research laboratory (International Institute for Strategic Studies, 2018), confirming in a press release that the new laboratory allowed the DSO to experiment with 'Unmanned Ground Vehicles and Unmanned Aerial Vehicles [that] work seamlessly as a team, without heavy reliance on human operators' (Ministry of Defence, 2017). While the majority of the DSO's work is related to defence, it does partner with civilian researchers and, somewhat more unusually, the DSO has previously bid for corporate research funding where there are potential military applications (Markowski et al., 2009).

The second research agency, the Future Systems and Technology Directorate, focuses on injecting strategic perspective into the development of SAF doctrine. The FSTD was originally created in 2013 from the merger of the Future Systems Directorate and the Defence Research and Technology Office (Laksmana, 2017), and operates under a greater level of secrecy than the other defence research agencies. FSTD has been credited with major roles in the development of the Advanced Combat Man System, which leverages wearable communication and information technology to improve the lethality and survivability of individual soldiers, and the Airspace Management Technology system, which regulates the busy Singaporean airspace. The FSTD also analyses emerging operational concepts globally to glean strategic perspectives on the tenets of future warfare that could be applied to the SAF. For example, a 2015 speech by the defence minister linked autonomous military technology with the Next-Gen SAF (Raska, 2015), while behind the scenes the FSTD had been examining LAWS-related operational concepts, such as the US Third Offset Strategy. Overall, the main role of the FSTD is similar to that of the United Kingdom's Development, Concepts and Doctrine Centre.

As mentioned above, the Defence Science and Technology Agency is a corporatized state agency with responsibility for coordinating military research and development, while also advising on the SAF's procurement process. Similar to Indonesia's *Komite Kebijakan Industri Pertahanan*, the DSTA is semi-autonomous and chaired by the permanent secretary of the Ministry of Defence (Laksmana, 2017). In effect the DSTA operates as a kind of technocratic gatekeeper and plays a key role in the acquisition of weapon systems, mapping out required defence capabilities and implementing defence innovation plans. As an example, the DSTA was responsible for managing the acquisition of Heron 1 UAVs from Israel, as well as heading the subsequent process of upgrading the Heron's datalink system. The DSTA is structured around 18 programme centres, which are each comprised of key subject area clusters; for example, the Advanced Systems Programme Centre contains three clusters: communications, sensor technology and guided weapons. As early as 2007 the DSTA sponsored a series of competitions for 'urban warrior' robot designs that could participate non-lethally in counter-terrorism operations (Horowitz, 2014). In March 2017, this agency expanded to include a laboratory dedicated to analytics and artificial intelligence (International Institute for Strategic Studies, 2018). As the primary agency responsible for coordinating the SAF's innovation and procurement process, the DSTA would be expected to play a major role in the adoption of autonomous weapon systems by the SAF.

These military research agencies also work closely with the civilian research sector. Singapore's two leading universities, the National University of Singapore (NUS) and Nanyang Technological University (NTU), are both well-resourced and internationally respected. Domestically based think tanks like the S. Rajaratnam School of International Studies (situated in the Nanyang Technological University) play a major role in conducting security studies research and advising the Singaporean government. Singapore has consistently supported research and development partnerships between military research agencies, universities and corporate entities. For example, Singapore's National Science and Technology Plan (2001–05) included a commitment of USD 4 billion to transform Singapore into a 'knowledge-based economy' (Goldman and Mahnken, 2004). Over the period of 2000 to 2013, the Singaporean government committed to increasing its investment in research and development funding by an average annual rate of 6.8% (Hourihan and Parkes, 2016). Examples of these corporate–state research partnerships include when Singapore Technologies Engineering (STE) partnered with DSO National Laboratories and the NTU to establish ST Electronics (Satellite Systems), a partnership which, in 2015, launched Singapore's first commercial Near Equator Orbit Earth Observation Satellite. In another example, the SAF's first indigenous-built UAV (Skyblade III) was the product of a collaboration between DSO and Singapore

Technologies Aerospace (a private company). Finally, in 2017 STE announced a USD 150 million fund to support engagement with new start-ups working in robotics, autonomous system and data analytics technology (Grevatt, 2018). Of equal importance, it is clear that Singaporean companies and research centres have a greater level of meaningful integration into the SAF's innovation and procurement process than in Indonesia.

In addition to direct investment in new capabilities, the Singaporean Ministry of Defence allocates significant resources to what it refers to as 'investments in human capital' (Ministry of Defence, 2017), educating and developing the capabilities of its limited personnel. Reflecting the broader meritocratic approach to societal advancement in Singapore, young professionals and soldiers with identified potential are consciously groomed to assume high positions within civilian and military organisations. These soldiers and officers are supported to study at western universities (Laksmana, 2017) and undertake placements with allied militaries as part of the 'SAF overseas scholars framework' (Raska, 2015). Further, lucrative 'Dual Career' schemes are intended to improve retention of talented soldiers, while the SAF encourages older officers to retire in their fifties (Huang, 2009) to ensure space for the rapid advancement of these young officers (Laksmana, 2017). The introduction of the Enhanced Warrant Officers' Career Scheme and Military Domain Experts Scheme in 2009 was designed to retain and promote experienced, well-educated senior NCOs and subject matter experts, respectively. The SAF is clearly invested in developing and retaining its 'talented' officers as part of its effort to keep the permanent 'core' of its military young, engaged and well-educated.

However, Singapore, unlike its original partner Israel, has a strict social hierarchy that stifles experimentation and revolutionary innovation. This was apparent in the SAF's evolution, which has been exclusively the result of top-down directed innovation. Although its senior leadership since the 1990s has been exclusively higher educated, there is still a notable lack of meaningful debate within the SAF on its future strategic direction (Raska, 2015). The establishment of the Future Studies Directorate in 2004 was a promising sign that the SAF was interested in fostering the debate and disruptive experimentation advocated in the 'Creating the Capacity to Change' *Pointer* monograph (Choy et al., 2003; Tjin-Kai, 2012). In its first year, the FSD was allocated an estimated S$ 8.25 billion (roughly USD 4.842 billion in 2004 terms) and given the explicit task of acting as a 'stress-test' for the established strategy and operational assumptions (Raska, 2015). This remains the principal role of the post-merger Future Systems and Technology Directorate. Despite these advances, the SAF remains committed to its cautious, considered evaluation and procurement process.

7.2.3.3 Organisational age

The final Organisational Capacity variable identified in ACT is Organisational Age. As with the TNI, the nature of the domestic civil–military relationship is a more effective measure for this variable, than the length of time since the most recent loss in a major conflict. Unlike the TNI, however, the SAF has been influenced by recent combat experience in support of coalition forces in Iraq and Afghanistan. The close civil–military relationship between the governing PAP and the SAF's senior leadership should theoretically limit internecine rivalry, ensure consistent levels of funding and promote collaborative development (Laksmana, 2017). In effect, a close civil–military relationship would indicate that recent statements by Singaporean officials linking autonomous military technology and the fourth-generation SAF are useful indicators that the SAF will indeed invest in acquiring autonomous military technology.

At the core of Singapore's civil–military relationship is the Total Defence strategic framework. Originating in 1984 (Matthews and Yan, 2007) the Total Defence strategic framework consists of six mutually supportive domains which contribute to Singapore's security (Raska, 2015). The traditional 'military defence' domain refers to the SAF itself, as well as the broader defence technological community. The 'economic defence' domain recognises the link between maintaining a powerful, innovative economy and the ability of the SAF to maintain its deterrent capabilities, while also ensuring that the domestic economy maintains the ability to transition to a war-production footing in the event that deterrence fails. The 'Psychological' defence domain reflects the importance of building societal resilience and embedding a collective will to defend the Lion City into its multi-ethnic population, while the 'Social' domain refers to government efforts to assimilate this population into its particular meritocratic system. In addition to its security role, national service in the SAF plays an important secondary role of acting as a 'melting pot' to bridge Singapore's ethnic divisions and instil a sense of nationalistic commitment in its conscripts (Tan and Lew, 2017). 'Civil' defence refers to the need to secure Singapore's vital resources (such as food, water and fuel) and infrastructure in the event of a conflict (Matthews and Yan, 2007).

The final domain, 'Digital Defence', was introduced in Singapore's 2019 budget, and demonstrates a recognition of the need to protect Singapore's digital, as well as physical, infrastructure (Tan, 2019). This was the first time a new domain has been added to Total Defence since its inception, yet the announcement was made by the civilian Communications and Information Minister S. Iswaran.[6] It also dominated the Defence Minister's Total Defence Day speech (Baharudin, 2019). While the official announcement was held at a graduation ceremony at Fort Canning Green, neither minister's remarks prominently featured senior uniformed SAF officers. Instead, both speeches repeatedly emphasised the role of the general public in digital defence (Iswaran, 2019). This emphasises the core contribution of Total Defence to understanding Singapore's civil–military relations, that the civilian government is firmly in charge.

From its establishment the SAF has been firmly subservient to the civilian government, an oddity in the region. This relationship has its roots in the SAF's colonial roots and was reinforced by the 1967 'Code of Conduct for the Armed Forces', which imposed a strictly professional role for the military while promoting loyalty to the government. In announcing this code, the incumbent Secretary of Defence, Goh Keng Swee, noted that:

Members of the SAF have a unique role; not only the ever-vigilant guardian of our nation but are also required to be an example of good citizenship. (Tan and Lew, 2017)

This code eventually formed the basis of the SAF's formalised seven core values, introduced in 1996 ('Safety' was added in 2016). Combined with the fact that the early leadership of the SAF was primarily composed of civilians (Laksmana, 2017), these factors have led to a modern SAF whose officers simply do not have the same level of political autonomy as their Indonesian peers. Raska (2015) even argued that the modern SAF has essentially become an 'incubator for future public servants and industry leaders', while Laksmana (2017) has referred to early SAF officers as effectively 'civil servants in uniform'.

Singapore's 'unified' civil–military relationship and the dominance of civil authority is best illustrated by comparison to the TNI. Singapore's 'dual career' system, which allows military officers to concurrently hold positions within the civilian administration, reinforced civil control over the SAF. Compare this to the ongoing difficulty the Indonesian government has encountered in its attempts to implement the Minimum Essential Force strategic framework where it conflicts with the Total People Defence System, despite the TNI's official disapproval of military personnel participating in political decision-making.

The unified nature of Singapore's civil–military framework has embedded an institutionalised approach to innovation that prioritises a cautious, sober, comprehensive assessment and procurement process for new military platforms. While this process can inhibit disruptive or revolutionary innovation, it has ensured the ability of the SAF to commit to long-term 'spiral' evolutionary innovation along the lines of identified strategic need. The close civil–military relationship lends credence to utilising speeches and policy statements by senior politicians (Raska, 2015) as indicative that increasingly autonomous military technology has been identified as a strategic priority that the SAF can evolve towards.[7]

7.2.4 Receptiveness of domestic audience

There is a complete lack of published data on whether the Singaporean public would support the development and deployment of autonomous weapon systems. Neither of the two surveys commissioned by the Campaign to Stop Killer Robots (2017 and 2019) nor the 2015 Open Roboethics Institute survey included Singaporean respondents. Despite this, the general consensus is that the population would support the pursuit of unmanned plans by the SAF and the Ministry of Home Affairs. This is supported by the fact that the Singaporean population is highly educated and generally supportive of emerging technology (Desker and Bitzinger, 2016). Historically the PAP has been able to legitimise the Ministry of Defence committing significant resources towards procuring expensive, advanced military platforms (Laksmana, 2017), and the population has seemingly embraced the widespread development of unmanned platforms and robotics in the commercial sphere (Desker and Bitzinger, 2016).

As of mid-2019 the Singaporean government has not explicitly stated its view on the permissibility of LAWS or whether it supports the ongoing calls for a pre-emptive development ban. Given that Singapore is not a signatory to the underlying convention, it is perhaps unsurprising that Singapore has not sent an official delegation to participate in the CCW Group of Governmental Experts on LAWS meetings, although Dr Collin Koh Swee Lean (a research fellow at the S. Rajaratnam School of International Studies) presented a research paper in his personal capacity at the 2016 meeting (Lean, 2016). Singapore eventually sent an observer detachment to the 2018 Meeting of High Contracting Parties to the Convention on Certain Conventional Weapons, although autonomous weapons were not the only issue under discussion and the delegation's specific contribution to the meeting was not publicly noted.

Fortunately, as with Indonesia, Singapore is a member of the Non-Aligned Movement, which made public statements on autonomous weapon systems in 2017 and 2018. These statements were considered in the preceding chapter. As with Indonesia, the importance NAM places on the principle of non-interference in the domestic affairs of member states suggests that these statements are more useful as guidance than as a definitive reflection of Singapore's position on LAWS and other autonomous military technology.

A key aspect of the dolphin strategy is promoting and participating in cooperative regional security efforts. This can be seen in Singapore's post-2015 approach to counter-piracy and in its contribution to nuclear non-proliferation discussions at the 7758th meeting of the UN Security Council in August 2016. It is therefore notable that Singapore has not adopted a similar, public stance supporting regional cooperation to respond to the emerging issue of autonomous weapon systems.

On the domestic front there is significant evidence that Singapore is considering the benefits of adopting autonomous military technology. In addition to the RSN's active participation in the ongoing development of the Protector Class USV following its first deployment in 2005, senior civilian officials have made public statements expressing interest in autonomous weapon systems. This support was reflected in the allocation of additional funding in 2017 for the DSO and DSTA to establish artificial intelligence and robotics research laboratories (International Institute for Strategic Studies, 2018).

While it is unfortunate that quantitative data on the Singaporean public's opinion of autonomous military technology has not yet been gathered or published, in the case of Singapore this information would not be as impactful as, for example, if examining Australia or New Zealand. This is because the electorally dominant PAP has historically demonstrated a capacity to legitimise remarkably generous allocations of the Lion City's limited resources to the procurement of advanced weapon platforms under the banner of defending Singapore's security and sovereignty. While this has been challenged by the opposition somewhat in recent years, the military's revised budget remains at 3–4% of GDP, significantly above Indonesia or Australia. Further, as discussed above, the percentage of military spending earmarked for research, development and procurement is routinely shrouded from public view. This creates an environment where the SAF could conceivably develop and procure increasingly autonomous military technology without significant political opposition. Furthermore, as noted by Desker and Bitzinger, if Singapore was to be threatened or attacked by an aggressor state, the existential threat posed to the city-state would render any political concern about the use of armed unmanned platforms inconsequential in the minds of Singapore's leadership (Desker and Bitzinger, 2016).

7.2.5 Capacity to develop or emulate specialised operational praxis

The final diffusion variable to consider is whether the SAF has demonstrated the capacity to develop or emulate a specialised operational praxis for the deployment of autonomous weapon systems. An operational praxis is the process through which a military transforms capability into force and is therefore a key factor in determining how a state engages with the emergence of a military innovation. While the prior sections of this chapter have demonstrated that Singapore has a significant adoption capacity, relative to its status as a middle power and compared to its regional neighbours, this section will explore whether the SAF has the capacity to overcome its structural preference for evolutionary, 'spiral development' rather than disruptive change to its strategic thought (Raska, 2015). This is particularly important within the context of the planned Next-Gen SAF transition, which prominently features autonomous systems and artificial intelligence-supported capabilities in its planned replacement of most major platforms by 2030 (Wong, 2019).

A prominent aspect of the SAF's evolutionary form of innovation has been emulating the operational concepts developed by larger military powers, albeit informed by a level of domestic modification. This was apparent in all three prior generations, beginning with the influence of the Israeli ministry on the early structure of the SAF. In this case, Singaporean officials integrated Israeli platforms and organisational structures while strongly distancing the SAF from the IDF's aggressive stance of forward-leaning deterrence.

This continued with the development of the 3G strategic concept in the 1990s and early 2000s, which prominently drew on observations of the US military's application of network-based warfare in the Gulf War and Kosovo. Raska (2015) highlights a monograph published in *Pointer* (the SAF journal), which outlined how the SAF could modify and incorporate network-centric warfare from the United States into what became the Integrated Knowledge-based Command and Control (IKC2) doctrine. The IKC2 monograph further argued that the SAF needed to combine the SAF's top-down model with disruptive, bottom-up innovation. This built on an earlier *Pointer* monograph, titled 'Creating the Capacity to Change', which had argued that the SAF needed to assume that disruptive innovations would emerge and therefore needed to develop an effective capacity to rapidly evaluate and potentially adopt these disruptive weapon systems (Choy et al., 2003).

While the SAF moved beyond simple replication in the application of these observations by the time the 3G SAF operational umbrella was announced in 2004

(Laksmana, 2017), the third generation of Singapore's military still utilised modified versions of operational praxes that were originally developed by the United States. A practical comparative analogy can be drawn to the Singapore Ministry of Defence's dual-acquisition process, whereby niche indigenous capability is leveraged to upgrade and modify advanced platforms procured from external allies, such as the F-5E Tiger II aircraft, originally purchased from Northrop Grumman, which underwent a domestic upgrade of its armament and navigation systems (Matthews and Yan, 2007). While emulating first movers has advantages for fast followers in the diffusion of an innovation, the SAF's structured, top-down modernisation process has historically stifled its capacity to successfully adopt more disruptive doctrinal innovations.

7.3 Evaluating Singaporean response options

As outlined in the methodology chapter, there are five response options available to states following the demonstration point of Lethal Autonomous Weapon Systems. While they are not necessarily mutually exclusive, determining the 'correct' response, or even the most likely combination to succeed, for Singapore is largely dependent on its adoption capacity, the threat environment, and the actions of neighbouring state and non-state actors.

7.3.1 Limited adoption

Singapore's interest in autonomous military technology was reflected in the Next-Gen SAF strategic framework, which was publicly announced in 2019. This interest is not surprising given that Singapore's defence policy has consistently prioritised the maintenance of a credible deterrence capability through a technological offset comparative to its neighbours. While Singapore does not currently have the adoption capacity required to adopt LAWS of the sophistication or capability level pursued by major powers, it would likely be able to successfully adopt less advanced platforms. Furthermore, given the prior emulation and incorporation of elements of networked warfare into the 3G SAF strategic framework, it is likely that the SAF would be able to draw on operational praxes and concepts developed by other militaries to successfully integrate these platforms. Based on an application of the adoption variables above, Singapore appears to be following an established pattern of careful review and slow adoption in response to an identified capability gap or emergent military technology.

Examining the, admittedly limited, publicly available information on the planned Next-Gen SAF indicates that the SAF continues to draw lessons from the practices of advanced military allies, particularly in relation to the deployment of increasingly autonomous military technology. The Next-Gen SAF strategy outlines an intention to equip infantry with low endurance short-range UAVs for tactical situation awareness (Wong, 2019). This use of unmanned aircraft has emerged as the most common operational praxis for the deployment of remote-operated and semi-autonomous, low-cost UAVs. While the SAF is developing a new medium-endurance UAV for surveillance, it is not expected to adopt a weaponised version. Interestingly, ST Kinetics announced in March 2018 that they were developing an armed close-range quadrotor UAV, the Stinger Unmanned Aerial Multi-Rotor Gunship, which would be included under the Next-Gen SAF strategic umbrella. The Stinger is armed with a light machine gun and intended to provide fire support role for company-level infantry units. Tellingly, ST Kinetics is developing an 'assisted threat identification function', whereby 'all a soldier needs to do is to designate the threats that need to be neutralised and the Stinger will automatically persecute the selected targets' (Wong, 2018).

The Republic of Singapore Air Force is also reportedly developing the capability to deploy unmanned aircraft to autonomously identify damage to runways and military installations, further reducing operational costs. This is a continuation of the SAF's longstanding reluctance to adopt weapon systems that would be seen as aggressive by its neighbours, although as stated above, Singapore is likely to be developing the capability to rapidly introduce an armed variant of these platforms in the event of hostilities (Zhang, 2018). In light of Singapore's apparent disinterest in medium and long endurance strike-capable UAVs it is worth highlighting that Singapore would not require strategic level autonomous weapon platforms to pursue their critical task focus, and that this is already how they perceive autonomous weapon systems.

The SAF's use of increasingly autonomous military technology in the ground domain is also developing largely in line with that of advanced powers such as the United Kingdom. Here the SAF demonstrates a similar lack of interest in overtly aggressive platforms, such as the US Sea Hunter or the Russian Uran-9. The SAF is known to be developing unmanned watchtowers with three deployed to date. These towers take advantage of the immunity of autonomous platforms to poor weather and fatigue to reduce the manpower required by a third (Zhang, 2018). These towers also have the additional advantage of being re-deployable within the confines of Singapore (Singapore Ministry of Defence, 2018). Singapore has also expressed an interest in tactical-level weaponised unmanned ground vehicles, with ST Kinetics announcing the development of the Remote Weaponised Soldier-class Unmanned Ground Vehicle. ST Kinetics had earlier unveiled a version of the Probot (originally Israeli made) equipped with an ADDER Remote Weapon System at the 2018 Singapore Air Show. Adopting a similar operational praxis to its allies, these UGVs appear to be intended for deployment in support of human soldiers and operate in a supervised or semi-autonomous role.

In the maritime domain, emerging operational doctrine clearly reflects the SAF's critical task focus on defending Singaporean waters and interests within the context of regional diplomacy and security cooperation. This places the acquisition of unmanned surface and underwater vehicles clearly within the conceptual framework of the dolphin strategy, developing capability to respond efficiently to regional security threats without adopting aggressive or threatening platforms. As a comparison, consider the recent decision by South Korea to name its new class of amphibious assault ship after the Dokdo island chain, ownership of which is disputed with Japan (Farley, 2018). Adopting, and subsequently arming, platforms like the Protector USV reflect a commitment to maintaining security in and around Singapore, which is sensible given the economic risk posed by piracy and maritime terrorism. The Next-Gen SAF strategic umbrella goes a step further with the development of Multi-Role Combat Vessels, which have been described as essentially 'modular motherships' for unmanned platforms (Wong, 2019). The mothership operational concept is being actively pursued by the United States and China; however, Singapore would arguably gain significantly more from its implementation. This would allow the Singaporean Navy to more effectively coordinate regional maritime security patrols and improve the 'eyes in the sky' capability that was so crucial to its post-2015 success. As with remote-operated platforms, autonomous platforms offer less political and operational risk and can therefore be deployed in riskier or politically fraught situations with less chance of sparking unintentional hostilities. Furthermore, the MRCVs will reportedly also be significantly automated to reduce manpower and other operational costs by the time of their expected deployment in 2030 (Wong, 2019).

Finally, from an industrial standpoint, the SAF has indicated a similar interest to its European allies and the United States in developing unmanned systems for performing surveillance and maintenance tasks on aircraft. Continuing the trend of partnering with commercial companies to bridge capability gaps, the DSTA signed separate agreements with Airbus and Boeing in 2018 to cooperate on research and development efforts aimed at integrating autonomous systems into future Singaporean airbases in security, surveillance and maintenance roles (Grevatt, 2018). While there have been no reported efforts, it is likely that the Singaporean defence technology community is developing, or at least considering, emulating the British operational concept of deploying artificial assistants to improve the efficiency and reduce the vulnerability of command posts. Finally, Singapore's intended deployments of limited autonomy platforms in these manners would be an effective offset for the increasing challenges it faces with an ageing population; however, it are insufficiently advanced or disruptive to prevent regional rivals from adopting similar capabilities.

Overall it seems clear that the SAF's interest in adopting autonomous platforms will reflect aspects of its adoption capacity, chiefly the Next-Gen SAF includes operational praxes that reflect Singapore's preoccupation with the threat of terrorism, its need to maintain regional security cooperation in the light of its thinning resource advantage over other key ASEAN states and the increasing pressures of maintaining a suitable sized defence force in the face of a rapidly ageing population that was already significantly smaller than its neighbours. It is, therefore, far more likely that the Singapore Armed Forces would limit its attempts to adopt autonomous systems to enhancing and augmenting traditional capabilities, while utilising AI-enabled systems to reduce the crew requirements of platforms and replace human soldiers in dull, dirty or dangerous logistical and surveillance functions. It is highly unlikely that the SAF would attempt to develop long-range or purely offensive systems, as this would be likely to increase the security dilemma of Singapore's neighbours, and thus would be seen as counter-productive by the Ministry of Defence.

7.3.2 *Counter-innovation/offset*

The SAF has already demonstrated a willingness to invest in capabilities that could partially offset the advantage gained by a neighbouring state adopting autonomous military technology. Two of the current potential counter-innovations to LAWS are cyberwarfare and remote-operated weapon platforms. As with Indonesia, adopting a counter-innovation strategy would allow Singapore to limit the impact of a rival adopting autonomous weapon systems as part of an offset strategy or the increased regional influence of a major power adopter.

As outlined in Chapter 4, Singapore has invested heavily in developing, procuring and deploying remote-operated platforms under the 3G SAF strategic framework. Remote-operated platforms provide the SAF with some of the key benefits of autonomous weapon systems without the potential diplomatic risks of attempting to adopt LAWS while the CCW debate is ongoing. These benefits include limiting the impact of its ageing population, increasing the combat effectiveness of its comparatively small military and acting as a force multiplier for surveillance efforts. However, limiting itself to less-advanced remote-operated platforms would be contrary to the SAF's longstanding determination to maintain a strategic offset, which it views as critical for maintaining credible deterrence (Desker and Bitzinger, 2016).

Befitting a technologically advanced, internationalised economy, Singapore has strongly committed to improving its offensive and defensive capabilities in the emerging cyber domain. The risk of a major cyberattack was recognised by Singapore's security forces during the transition period towards the 3G SAF, with the Cyber Defence Operations Hub established from a merger of existing cyber operations units in 2013 (Raska, 2015). In 2017 the SAF established the Defence Cyber Organisation as a distinct strategic command, responsible for implementing cyber security policy and defending Singapore's military cyberinfrastructure (International Institute for Strategic Studies, 2018). The DCO is well resourced with a full establishment of 2,600 personnel spread across four formations, each commanded by a flag-rank officer or equivalent civilian official. During the same period, the SAF established the Cyber Defence Group, which has responsibility for protecting the SAF's networks and providing incident response (International Institute for Strategic Studies, 2019). More recently, Singapore re-emphasised its commitment to cyber operations in 2019 when it took the unprecedented step of adding cyber defence as the sixth pillar of the 'Total Defence' framework, which also involved establishing the Home Team Science and Technology Agency under the auspices of the Ministry of Home Affairs (Tan, 2019).

Overall it is apparent that Singapore has already begun to invest in developing a significantly greater capacity than Indonesia to utilise either remote-operated platforms or cyber operations. While this could be relied upon to partially offset the impact of a neighbouring state adopting autonomous weapon platforms, given its broader adoption capacity and critical task focus, the SAF would be better served by integrating these capabilities into an adoption-based response.

7.3.3 Balance

An important caveat to Singapore's stated preference for maintaining armed neutrality is that it has always been tempered by the recognition that strong multilateral cooperation is vital to maintaining regional stability and offsetting the city-state's inherent vulnerability. Singapore's leadership recognises that self-sufficient deterrent capability must be partnered with regional and global diplomatic efforts to create the 'political, diplomatic and economic space' (Tan, 2015) the city-state needed for growth and development beyond simple survival. As a middle power state surrounded by much larger states (in geographic and population terms), Singapore is strongly incentivised to promote adherence to international laws and norms as a way to protect its interests and ensure much-needed regional stability (Chan, 2016).

Singapore's membership in ASEAN is a lever through which has been able to exert an outsized influence on its neighbours relative to its size. For example, Singapore is a major coordinator of regional counter-piracy efforts as the host of ReCAAP and hosts the influential yet unofficial Shangri-La Dialogue (Tan, 2015). Furthermore, in a similar manner to Indonesia, Singapore has been positioning itself as a trusted intermediary within ASEAN for negotiations with China, particularly over the South China Sea dispute (Australia-China Relations Institute, 2015). Singapore has a strong interest in ensuring that the ASEAN member states retain a cohesive, cooperative approach towards China and the United States, especially in the event of deepening hegemonic tensions. This was highlighted in 2016 when two Singaporean diplomats publicly accused China of attempting to interfere in the internal decision making of ASEAN member states (Fook, 2018). Singapore's desire to balance its global and regional relationships has historically caused tensions with other ASEAN members (Tan, 2015). More recently, its approach to both the Chinese One Belt One Road Initiative and the US pursuit of increased tariffs against China has been influenced by the broader ASEAN viewpoint.

Singapore's response to the emergence of autonomous weapon systems will be influenced by its fellow ASEAN member states, particularly Malaysia and Indonesia. It is therefore likely that the SAF's approach to integrating increasingly autonomous military technology and any future adoption of autonomous weapon systems would be shaped by the need to balance not only Singapore's political and security interests, but also how that adoption would be perceived by its neighbours. However, it must be noted that survival dominates Singaporean strategic thinking, and this will influence how the SAF approaches AWS. Given that autonomous military technology has been identified as a priority by the defence technological community and that LAWS are being trumpeted as a major factor in future warfare, it is highly unlikely that Singapore would willingly participate in any international effort to limit its access to this technology. This can already be seen in its reluctance to participate, even informally, in the CCW process and must be considered part of its likely response to a LAWS demonstration point.

7.3.4 Bandwagon

While this would in theory be a justifiable response option for a middle power state with a long-standing defence relationship with a major power, Singapore would be poorly served by bandwagoning because its continued economic growth and security is reliant on the maintenance of a stable balance of power in the region. It appears far more likely that Singapore would prefer to continue to carefully balance its relationships with China and the United States following the demonstration point of LAWS.

Singapore is one of the closest US security partners in Southeast Asia and its leaders have repeatedly expressed the view that the superpower acts as a stabilising force in the region and globally (Dexian, 2013). Singapore conducts major joint-exercises with the US military and actively participates in officer exchanges. The SAF has also purchased multiple advanced platforms historically, and in 2019, became the first Southeast Asian state (Panda, 2018) to purchase four F-35 Joint Strike Fighters (Singapore Ministry of Defence, 2019). While Singapore has exhibited at least a tacit acceptance of the United States as the dominant hegemonic power and remains closely linked in security terms, its leadership has been careful to publicly maintain a level of policy independence. Recent examples include Singapore's decision to continue with the Trans-Pacific Partnership despite the withdrawal of the United States, and the city-state's continued refusal to explicitly support the controversial freedom of navigation patrol operations in the South China Sea despite allowing the US Navy to resupply its Carrier Battle Groups at the Changi Naval Base since 2000 (Dexian, 2013). Despite its public stance, Singapore has still been criticised for supporting the United States in its efforts to balance the rise of China, particularly in the aftermath of the 2016 pivot, which senior Singaporean policymakers praised in public speeches (Fook, 2018).

As with other states in Southeast Asia, Singapore is faced with the challenge of balancing its political and security ties with the United States with its deep economic relationship with China. In 2009 Singapore became the first Asian state to successfully seal a bilateral free trade agreement with China and, by 2013, China had overtaken Malaysia to become Singapore's largest bilateral trade partner, a position it continued to hold as of 2019. This trade relationship is clearly on track to expand under China's One Belt, One Road Initiative and Singapore was one of the first states to join the Asian Infrastructure Investment Bank, despite opposition from the United States.

However, Singapore has also demonstrated caution towards China. While the city-state is not directly threatened by Chinese military modernisation and is not a claimant in the South China Sea disputes, its preference for multilateral engagement with China through ASEAN demonstrates a recognition of the importance of international norms and law for preserving the city-state as a middle power. This directly led to a deterioration in Singapore–Chinese relations in 2016 when the two states were entangled in a public dispute over Singapore's perceived role in ASEAN attempts to shift the language in a Non-Aligned Movement statement on the South China Sea, while Singaporean diplomats accused China of attempting to exercise undue influence on individual ASEAN member states (Fook, 2018). Furthermore, Singapore's relationship with Taiwan, while unofficial due to its public subscription to the One China policy, irregularly causes tension with China, which views Taiwan as a renegade province (Dexian, 2013). A final reason that Singapore would be reluctant to explicitly join with China as an early adopter of autonomous weapon systems is rooted in local diplomacy. Singapore has made considerable efforts to offset the perception that its ethnically Chinese majority population makes the city-state a 'Chinese island in a Malay sea'. Bandwagoning with China would undermine this effort and increase tensions with Indonesia and Malaysia, which would in turn damage Singapore's security, as well as its efforts to enlist Southeast Asian states into broader regional security cooperation.

Of the three external response options, Singapore is the least likely to pursue a response strategy that prioritises bandwagoning with an early adopting great power state in the region. Aligning itself too closely with either the United States or China in the event of increased hegemonic tension following the demonstration of LAWS risks alienating the other power as well as valuable potential allies among the other ASEAN and Five Power Defence Arrangement states in the region. It is far more likely that Singapore will avoid this response option as part of its traditional guarded neutrality.

7.3.5 Reassert neutrality

Based on its historic diplomatic stance and current defence framework it is clear that Singapore will endeavour to maintain its guarded neutrality as the core of its response to the emergence of Lethal Autonomous Weapon Systems. As discussed earlier, maintaining an 'ideology-free' approach to international relations and avoiding restrictive military alliances has always been central to Singaporean foreign policy (Tan, 2015). The Lion City would be poised to benefit significantly if it is able to balance its commitments to both major powers during a transition. Tellingly, Singapore has remained at a distance from the ongoing negotiations surrounding the benefits of a developmental ban under internal law, engaging with the Convention on Conventional Weapons primarily through the Non-Aligned Movement. While it is unlikely that Singaporean policymakers would depart from this longstanding foreign policy approach, neutrality does not necessarily mean inactivity.

Rather Singapore's approach will likely reflect the dolphin strategy, blending diplomacy with deterrence capability. Therefore, while reasserting neutrality would be a crucial component of its response, it would be most effective alongside a sustained, and likely successful, attempt to adopt a level of autonomous military technology. The SAF has already outlined its commitment to pursuing unmanned and increasingly autonomous military technology as part of its Next-Gen SAF framework (Wong, 2019).

Overall, therefore Singapore is likely to reassert its neutrality and attempt to balance its commitments to regional and global actors in the immediate post demonstration point period, however, this will be only a part of its overall response effort. Singaporean policymakers have repeatedly stressed that they have no perpetual enemies (Dexian, 2013) and that they are committed to maintaining its status as a trusted partner, even if this results in 'slightly warmer soup with either China or the US' (Fook, 2018). However, in the event of a direct threat to Singapore it is likely that the SAF would prioritise the rapid adoption of autonomous weapon systems, and given its adoption capacity, this would likely be successful in the short term.

7.4 Conclusion

Although Singapore has a greater adoption capacity than Indonesia, it remains a middle power state and would therefore be unable to compete against a great power state to become the first mover. However, Singapore is well placed among ASEAN member states to successfully undertake limited adoption as part of a broader response to the emergence of Lethal Autonomous Weapon Systems. The SAF has already committed significant resources to developing or purchasing unmanned platforms and increasingly autonomous military technology as part of the Next-Gen SAF strategic framework. While its preference for slow, carefully reviewed weapon procurement processes may delay the SAF, the continued investment in the FTSD reflects a renewed interest at the command level in developing disruptive doctrinal innovations, which is promising for Singapore's capacity to successfully integrate autonomous weapon systems.

However, this must be complemented by a reassertion of Singapore's neutrality between the great powers and a continued pursuit of regional security cooperation to increase stability, reduce intraregional tensions and counteract the threat of terrorism. Furthermore, the SAF's adoption of autonomous weapon systems would be constrained and shaped by its critical task focus on defending Singapore and its interests within a 'smart power' framework. Therefore, it is far more likely that Singapore will acquire platforms with short to medium range for surveillance alongside a selection of platforms and systems to increase the capabilities of its forces, which are being drained by an ageing population. This would allow the SAF to continue to maintain its technological offset while not upsetting its balancing act towards China and the United States or, arguably even more crucially, sparking conflict with its neighbours.

Notes

1 Quoted in Tan (2015).
2 Interestingly, piracy trends reflect a preference for targeting commodities that are particularly valuable at a given time; for example, crude palm oil was favoured between 2001 and 2011 but has now been replaced by crude petroleum.
3 As outlined by Laksmana (2017), these stages were:

(1) acquire new equipment, introduce progressively more capable systems, and establish new units to enable SAF's transformation into an advanced, networked force; (2) set up new operational command relevant with an expanded spectrum of

operations and, in doing so, focus on widening its operational flexibility and responsiveness; and (3) aim on enhancing SAF's leadership and human capital through the introduction of enhanced career streams as well as revision of training and curriculum to maintain a steady stream of capable and committed officers.

4 Bin Osman, M. M. (2014). 'Speech by Minister of State for Defence, Dr Mohamad Maliki Bin Osman, at the Young Defence Scientists Programme Congress 2014', Ministry of Defence. quoted in Raska, M. (2015). 'Military innovation in small states: Creating a reverse asymmetry'. London: Routledge.
5 Only seven of the DSO National Laboratories divisions are research focused, the other three focus on quality assurance, human resources and corporate services.
6 Minister Iswaran is also Singapore's minister-in-charge of cyber security.
7 This interpretation was also supported in personal communication between the author and a senior Singaporean defence analyst.

8 Discussing the impact of AWS diffusion on relations of power and strategic stability in Southeast Asia

8.1 Introduction

In the modern geostrategic climate, the Southeast Asian region hosts some of the most important and concerning potential flashpoints for inter- and intra-state conflict. The risk of conflict and competition in the region is combined with immense economic potential and an enduring level of intra-regional suspicion, even between ASEAN member states. The result is a region where states are developing towards middle power status, with the economic and political growth that entails under the shadow of ongoing hegemonic tensions between an existing superpower and a rapidly strengthening rival. It is into this environment that unmanned military platforms have already begun to proliferate along a path that will also be followed by the first generation of lethal autonomous weapon systems.

Prior military diffusion analyses have generally focused on great power states, proceeding on the implicit assumption that small and middle powers would be unable to effectively adopt the major military innovation and thus be forced to align with a great power competitor to preserve their comparative status during the hegemonic transition. While conflict and power rebalancing could occur on the regional stage, this has been generally subordinated to the broader hegemonic conflict spurred on by the emergence of a disruptive military innovation.

This book has challenged this approach, demonstrating that Southeast Asian middle power states have the capacity to effectively pursue, acquire, develop and/ or adopt unmanned military technology of varying levels of autonomy in both armed and unarmed variants. Therefore, the purpose of this chapter is to evaluate how the emergence of LAWS would impact the balance of power within Southeast Asia and to demonstrate how early adoption by middle power states would challenge prior conceptions of the hegemonic power competition and conflict. There are two sections within this chapter, contributing to both the analytical and theoretical aspects of this book.

The first section of this chapter focuses on evaluating impact at the regional level, based on the response options analysed in the previous chapter. This section argues that the proliferation of increasingly autonomous weapon systems will exacerbate existing tensions between ASEAN member states, which mistrust one another and place a high value on security and sovereignty, particularly as

DOI: 10.4324/9781003172987-8

deepening hegemonic competition erodes confidence in the liberal rules-based international order. Historically, more advanced middle powers in the broader region (including Singapore and Australia) have been determined to maintain a knowledge edge over their larger, but less advanced neighbours. When one considers the dual-use nature of key enabling technologies, the rising purchasing power of less advanced Southeast Asian states and the expected decrease in the unit cost of unmanned platforms over time, it becomes apparent that this is not an innovation that states will be able to limit their neighbours' access to. This means that any knowledge edge or capability offset achieved by a state like Singapore would only be transient. This will in turn incentivise states to regularly re-establish an offset with further incremental innovation, further raising the security dilemma of their neighbours and making intra-regional clashes and confrontational behaviour more likely in the absence of an effective and proactive response.

The second section of this chapter shifts its focus to an examination of how the adoption of increasingly autonomous weapon systems by middle power states challenges existing theories of hegemonic power transition, competition and conflict at the super-regional level. The capacity for non-great power states to become effective early adopters of this military innovation represents a unique opportunity for global southern states to gain and retain a greater level of independence from the competing states during hegemonic competition. Rather than transitioning through a period of bipolarity towards either a new hegemonic state or a re-assertion of the existing international order, the levelling effect of increasingly autonomous weapon systems suggests a return to a multipolar competition space, where influential regional actors are less subordinated to one of the hegemonic camps. This section concludes by engaging with the question of whether the Thucydides Trap is the incorrect lens for the emerging hegemonic conflict between China and the United States, proposing instead that the proliferation of AWS will instead lead to an arming of the modern Melians.

Overall, this chapter engages directly with the core research puzzle, arguing that, without proactive and regionally shaped action, the diffusion of increasingly autonomous weapon systems to states in Southeast Asia will negatively impact security and stability at the regional and super-regional levels. This chapter will draw on analysis from across the book to evaluate how the diffusion of LAWS (or a derivative) to middle power states would impact a future hegemonic conflict and the imposition of a new balance of power by the victorious hegemonic power.

8.2 Regional security impacts of AWS proliferation in SE Asia

Beginning with a regional security perspective, the development of increasingly autonomous systems in the absence of effective regulation presents policymakers with both challenges and opportunities. The main security challenge is structural; Southeast Asia is characterised by historical tensions and mistrust between states that, while of varying internal stability, still place a high value on their own sovereignty and security. The emergence of a major military innovation in this regional context without an effective control framework creates a greater risk of conflict by lowering the traditional barriers and risks associated with confrontational state behaviour, while simultaneously increasing the likelihood of unintended or uncontrolled escalation.

However, this innovation also has significant potential military and regional security benefits that deserve consideration (Anderson, 2016). This in no way diminishes the potential of LAWS to instigate disruptive change in the nature and patterns of international security and conflict, nor would their use lead to any form of 'sterile' warfare where human suffering disappears from conflict. Rather, it argues that under a recognised framework for their use, increasingly autonomous unmanned platforms could make a positive contribution to stability in Southeast Asia. Firstly, AWS are significantly more resource efficient at maintaining surveillance over remote or difficult-to-access regions. Furthermore, they are not as politically sensitive as manned platforms and require a lower level of operational security, enabling a greater level of multilateral cooperative deployment. To do so, however, willing ASEAN states must first adopt a common definition of unmanned platforms and establish an information-based normative framework for their usage, a step that is explored in-depth in the next chapter. Overall, the purpose of this section is to demonstrate how the demonstration and subsequent proliferation of increasingly autonomous weapon systems would impact the balance of power, regional stability and security in Southeast Asia.

8.2.1 Security dilemma, proliferation and the potential for a LAWS arms race

Understanding the regional security impacts of Indonesia and Singapore's expected responses to a future demonstration point of LAWS must begin with an acknowledgement that inter-state relations under an anarchical system, particularly in a region like Southeast Asia, operate on a foundation of perceived and projected power. If the balance of power in their region was to shift in favour of a neighbour, the other states would be incentivised to attempt to achieve a comparable increase in capacity to ensure their survival and continued influence, particularly given that states can never rely on having perfect information as to the intention of their neighbours (Fels, 2017). Due to their impact on the power projection paradigm, the emergence of a major military innovation is disruptive to established balances of power. States are thus forced to respond in order to preserve their status and survival, particularly when the innovation enables a successful hegemonic challenge. While there is concern evident in the evolving discussions and literature surrounding LAWS of the potential for this security dilemma spiral to devolve into an arms race, it is generally focused, somewhat understandably, upon great powers such as China, Russia and the United States. However, when one considers the diffusion of increasingly autonomous weapon systems into Southeast Asia, it is equally important to account for ASEAN member states.

For some states the adoption of a major military innovation is intended to be central to an offset strategy, which involves capitalising on a technical or operational advantage to artificially disrupt a disadvantage relative to a potential rival(s) and/or to gain an advantage by negating their dominance in a capability. Returning to the case studies, maintaining a technology-based offset is central to Singapore's security posture. The SAF is prominently pursuing unmanned platforms, increasingly autonomous weapon systems and artificial intelligence as part of its Next Gen SAF strategic concept, which was detailed in previous chapters. This is unsurprising given that the SAF's capacity to present a credible deterrent threat relies on leveraging a technological advantage to offset the disadvantages of Singapore's comparatively tiny population and complete lack of strategic depth.

Whether the disruptive advantage is technologically or operationally based, the effectiveness of an offset strategy relies upon maintaining the titular offset. Historically less advanced states were stymied from closing this capability gap by a number of barriers such as high resource capacity requirements, the need to access easily controlled components or reliance on highly specialised skill sets. These barriers slowed down the diffusion process or even limited the number of potential adopters, preserving the offset advantage. However, these barriers are substantially lower for entry-level unmanned systems due to their reliance on a dual-use enabling technology, although higher barriers remain to the adoption of advanced platforms. As demonstrated by the proliferation of remote-operated UAVs, as the unit cost of dual-use enabling technologies falls and capability improves, these barriers will continue to lower, placing additional pressure on states that hope to maintain a meaningful capability edge.

While there are no universally agreed definitional criteria for classifying proliferation as an 'arms race', this has not halted a steady stream of media articles suggesting that there is already an ongoing AWS or artificial intelligence arms race in progress and that their state is losing (Asaro, 2019). A commonality in most arms race definitions is the centrality of competition, that is to say that arms races involve a level of 'one-upmanship' as actors seek to secure or undermine comparative advantage through acquiring higher quality or more numerous platforms (Asaro, 2019). Huntington similarly placed the distinction on whether there was an 'absolute need' separate from bilateral competition, suggesting that some 'arms races' merely reflect modernisation efforts intended to provide an economic benefit to local industry (Horowitz, 2019). Indonesia's investment in modernising its domestic arms industry and Australia's current push to enter the top ten global arms exporting states are recent examples from the region of economically motivated decisions.

While the focus of prior studies of offset strategies has generally been on great power states, middle power states generally operate with a clearer picture of future adversaries and are thus more sensitive to military power shifts relative to potential rivals. This can be seen in Southeast Asia, where the underlying current of mistrust is being exacerbated by ongoing military modernisation efforts, which have become, to an extent, self-perpetuating (Wyatt and Galliott, 2018). In this environment, the defensive neorealist theory would suggest that AWS adoption (perceived or actual) by a Southeast Asian state would increase the security dilemma of its neighbours, incentivising neighbouring states to respond. This would in turn drive further evolutionary innovation and improvement by the original adopter, exacerbating the cyclical nature of security dilemmas (Fels, 2017).

Although this interaction between the security dilemmas of neighbouring states does not inevitably lead to war, it does escalate inter-state tension and raises the risk of conflict. Even if one accepts the neorealist premise that states are inherently rational actors, they are still prone to miscalculation, particularly when operating with limited knowledge, or a mistaken perception, of the intentions and capabilities of rival actors. This is particularly problematic with autonomous weapon systems because their operating software cannot be easily verified by neighbours, thus injecting even more uncertainty into assessments of the adopter's intentions and true capabilities (Horowitz, 2019). Overall, there is a significant risk that the adoption of AWS by a leading ASEAN member state in the current geopolitical environment would trigger a self-reinforcing process of cascading security balancing (including the adoption of increasingly autonomous weapon systems) in the region, which would make intra-regional clashes and confrontational behaviour more likely in the absence of an effective and proactive response.

8.2.2 Lowering barriers to warfare, provocation and unintentional conflict

A related concern is that the proliferation of autonomous systems would lower the perceived costs of warfare, prompting riskier displays of brinkmanship and provocation among states in the Asia Pacific, for example in cases of disputed maritime territory. The proliferation of autonomous weapon systems in this regional security environment raises the spectre of states being able to utilise force without the same level of consequence and with minimal political justification (Figueroa, 2018).

The availability of increasingly autonomous systems increases the risk of unintended conflict, provocation and escalation in three ways. The first is encapsulated in the argument that their development would contribute to a 'sterilisation' of warfare into a bloodless human-free form. The prospect of being able to coerce or impose force upon their neighbours without risking the lives of soldiers or the internal cost of an unpopular war is concerning because, regardless of its failings, this argument could be used to justify state decisions to resort to warfare without due consideration.

In a related manner, AWS proliferation would increase the risk of armed conflict because unmanned systems are inherently more deniable and expendable than their manned equivalents, which could encourage states into taking provocative actions (such as overflights of disputed islands) to make a political point, unrestricted by the risk of losing their own soldiers or expensive manned platforms. An aspect of this risk is uncertainty; in the absence of international law or visibly agreed-upon norms, it is not possible to know how Southeast Asian states would react to another state destroying or capturing one of its unmanned platforms in disputed territory. While recent cases between Turkey and Russia, and Iran and the United States did not provoke an armed response, there is no guarantee that this would be the case if, for example, Singapore shot down a TNI operated unmanned aircraft in murky circumstances. Furthermore, even when provocative actions between states do not result in a confrontation, they contribute to the reserve of bilateral tension and domestic pressure to preserve face in future disagreements, making the next provocation harder to peacefully diffuse (Cho, 2018).

Finally, the proliferation of AWS raises the risk of unintentional use of force due to system failure or mistake, potentially at speeds beyond human capacity to effectively intervene. While it is possible to deploy a weapon system today with full independent control over its critical functions, the deploying actor would have to accept a high rate of error and significant risk to civilians and friendly personnel in the area. Complex systems have a tendency to fail spectacularly and destructively, usually with a little obvious warning; AWS are no exception. Consider the following examples: early models of the SWORDS platform (an armed remote-operated UGV) possessed a glitch where they would suddenly spin on the spot, and in 2007 a South African anti-aircraft cannon malfunctioned and incorrectly engaged its own crew, killing nine soldiers and wounding 14 more (Shachtman, 2007).

Even if there is no technical error, a weapon system with autonomous control could engage a target that meets its criteria but that a human operator would have identified as an illegitimate or risky target. For example, in 1991 a Phalanx CIWS on the *USS Jarrett* misidentified chaff as a threat and fired on the neighbouring *USS Missouri*. Fortunately, no one was reportedly injured, principally because the *USS Missouri* was well outside the effective engagement range for the CIWS. While it was technically under human supervision, this incident highlights that autonomous systems do not have the capacity to subjectively analyse context outside of the available data in the same way that humans do.

When one further considers the phenomena of 'Flash Crashes' in civilian stock markets or the misidentification of a journalist as a terrorist facilitator by the Skynet meta-data analysis programme, the risk of an autonomously operating machine unexpectedly engaging a misidentified target or reacting with force to a deliberately provocative but merely demonstrative aerial intercept (as is regularly occurring between great power militaries in Europe and East Asia) becomes apparent.

A further risk centres on autonomous weapon systems interacting or detecting one another while patrolling in contested territory. If they were to engage there is no guarantee that the international community would ever be able to conclusively determine what led to that decision, and it is also feasible that this initial engagement would spread to involve nearby units (including human soldiers), which in turn could realistically escalate into an entirely unintended war between Southeast Asian states.

8.2.3 *Utilising AWS to respond to key regional security threats*

While being mindful not to diminish these regional security risks, and acknowledging that it is unlikely that leading ASEAN states (such as Singapore, Indonesia and Malaysia) will develop the level of C4ISR and data infrastructure (for transfer and storage) necessary to deploy unmanned platforms in the intercontinental manner of the United States, a more limited deployment of increasingly autonomous unmanned platforms could provide significant advantages over current, manned platforms. Furthermore, if properly monitored and regulated AWS could reduce the risks involved in peacekeeping operations to both civilians and soldiers, as well as offer states a lower-risk method for asserting claims in disputed territory than deploying a warship or coastguard vessel.

Arguably the most attractive advantage of utilising unmanned platforms is their increased resource efficiency, in other words, their lower cost per hour of operation. This is particularly attractive to Southeast Asian states, whose need for effective maritime domain awareness (MDA) is not reflected in regular or secure investment in maritime platforms and naval assets are commonly used in non-warfighting roles, including law enforcement (Laksmana, 2018). As an example, the Australian Strategic Policy Institute compared the cost-effectiveness of an MQ-4C Triton UAV and a P-8A Poseidon surveillance aircraft (based on the number of square kilometres covered per dollar of cost), which showed that the MQ-4C covers almost 10 square kilometres more per operations dollar than the P-8A (Mugg et al., 2016). Another highly efficient example is the Wave Glider, which can patrol a section of the ocean with visual and sonar sensors for up to a year, for a fraction of the cost of a manned patrol boat (Mugg et al., 2016). A UUV like the Wave Glider could identify suspicious shipping in remote waters and alert the regional information centre, allowing for a targeted response by manned or unmanned assets with less risk of unintentional escalation. States that want this capability without relying on expensive US military-grade systems could purchase Complete Off the Shelf (COTS) platforms from a myriad of state and civilian commercial providers.

Deploying a combination of unmanned platforms with varying levels of operational autonomy would provide a valuable contribution to regional efforts to reduce the ability of transnational organised criminal groups to operate in contested border waters. Because unmanned platforms do not require the same level of operational security and secrecy as manned platforms, they can be operated in a more transparent way. In this example, the ReCAAP fusion centre would be able to coordinate a targeted response by manned or unmanned assets, adding to the ability of neighbouring states to track illegal shipping and pirate vessels across international borders without the political cost of sending an armed military vessel.

8.3 Hegemonic power transition, competition and the Thucydides Trap

In the anarchic global environment, the emergence of a disruptive military innovation unlocks the potential for a challenger state to undermine the existing power superiority of a hegemonic power (in this case, the United States) by becoming a more effective adopter (either as a first mover or a superior fast follower). The attempt to close this power gap can build hegemonic tension during the innovation's incubation period, which can lead to conflict. The hegemonic state must maintain a sufficient capability edge to effectively project power in multiple regions while deterring challenges from near-peers (singularly or in alliance with smaller states). Conversely the challenger state views superior adoption and integration of the innovation as a way to undermine or offset the pre-existing power difference, enabling it to challenge the hegemon's role and increase its own. It is therefore important to understand how the diffusion of autonomous systems would impact great power competition in Southeast Asia as a regional hegemony.

Rather than focusing directly on great power competition this section adopts a Southeast Asian lens, recognising that a shift in the balance of power of this resource-rich and geopolitically influential region will have a direct impact on the broader Asia Pacific (Fels, 2017). Among the key goals of regional hegemons is to prevent other great powers from achieving too large of a role within their sphere of influence; however as Mearsheimer stated, they also have a vested interest in preventing rising powers from achieving dominance in neighbouring regions and may even intervene arbitrarily to support smaller state efforts to balance the rising great power. This can already be seen in Southeast Asia, where China and the United States have recently begun intermittent trade warfare and appear to be moving towards the Thucydides Trap as China attempts to minimise United States' regional influence in order to reassume its historical place as a regional hegemon.

Taking this a step further, the diffusion of AWS will place pressure upon great power states by forcing them to adopt or counter the disruptive military innovation based on imperfect information. Enabled by the dual-use nature of the underpinning technology, smaller states, presented with the same challenge of imminent instability, will imitate and emulate their larger peers as much as possible to secure their own power base (Goldman and Andres, 1999). Unlike in previous hegemonic transitions, the early proliferation of AWS to a greater number of actors will reduce the value of the US security guarantee and increase the risk of unexpected conflict within or between their coalitions (Nye, 2010).

8.3.1 What is the traditional role of middle powers in hegemonic power transition?

Limited by their comparatively minor national power generation and projection capacity, small-middle power states have historically lacked the capacity to compete with great power states. Horowitz points to the defeat of Belgium in 1940 as an example where, even with a 'perfect' response, the power differential between Belgium and Germany would have guaranteed their defeat. It is understandable, therefore, that neorealist accounts of hegemonic power transition focus on great powers. Without sufficient capacity to successfully attempt full-scale adoption of an emerging RMA, small and middle powers were incentivised to adopt external responses to protect their security and relative position.

Chief among these responses has been to join a balancing alliance or coalition against the first-mover adopter (or with, depending on the pre-existing relationship between a given state and the first mover). This offered smaller states protection and reduced uncertainty between lower-tier states, contributing to regional stability. The downside for smaller states was that their role in the subsequent hegemonic power competition was subordinated to the strategic interests of the coalition leader. The archetypal examples of coalition leaders in this respect would be Athens and Sparta. It is important to distinguish here that the interests of coalition members were historically subordinated, not necessarily subsumed, and advanced regional middle powers could still exert a level of influence within the coalition.

Regionally influential middle powers (such as Australia, Indonesia, South Korea or Singapore) leveraging their position to secure benefits from major powers support the application of a different reading of the Melian Dialogue, which is traditionally associated with various schools of Realism. A post-colonial security studies perspective on the dialogue highlights the fact that the Athenian siege eventually reached a stalemate. While objectively a weaker power, the effort to subjugate Melia consumed resources and political capital that could have otherwise been deployed against the Sparta-led coalition, and the Athenians eventually sought terms (Barkawi and Laffey, 2006). The lesson post-colonialist security scholars draw from this is that middle power states do not need to be able to win a war against a great power; they just have to be a sufficiently 'poisonous shrimp', to borrow the Singaporean imagery. Due to the opportunity cost associated with subjugating middle power states in the event of a hegemonic conflict, hegemonic competitors must combine coercion with an incentive to recruit and maintain their alliance network. Some states, particularly in Southeast Asia, have taken this a step further and attempted to balance their allegiance to both hegemonic competitors, aiming to secure support and connections from both.

However, maintaining a coalition of supporting states is also a valuable tool for a hegemon even without the threat of a rising competitor. Firstly, a strong alliance structure enables the hegemon to leverage the resources of other states and project its influence, while limiting the capacity of any rivals to develop competing influence in the region (for example, the post-Second World War US hub and spoke alliance model). Furthermore, even if the challenger secures a comparative bilateral advantage, the existing hegemon could draw on its stronger and more established alliance network for resources, development assistance, political support or military forces. This factor incentivises the challenger state to build its own coalition as well as to undermine the hegemon's perceived superiority and reliability among smaller states in order to reduce their commitment to the hegemon's coalition. To take a modern example, if Southeast Asian states no longer feel that they could rely on the US security guarantee, they could defect from the alliance, assert their neutrality or even take independent escalatory action.

8.3.2 *Competing for coalition influence – offset strategies and credible deterrence*

Maintaining, or conversely undermining, the support of Southeast Asian states for maintaining the United States' pre-eminence in the regional balance of power will therefore remain an important aspect of the emerging strategic competition between these two great power states. While the Trump administration's willingness to engage in intermittent exchanges of tariffs with China dominates the news cycle, China has already achieved a level of economic hegemony in Southeast Asia (Ikenberry, 2016). However, ASEAN states are wary of China's increasingly aggressive posture over the South China Sea territorial disputes (with the possible exception of Cambodia) and thus remain beholden to the security hegemony of the United States. However, the rise of autonomous weapon systems, which have been publicly identified as disruptive to the existing paradigm of conflict in which the United States is dominant, is injecting damaging uncertainty into the assumption of continued US primacy that underpins its hub and spoke alliance model (Le Thu, 2019).

From a geopolitical perspective, the United States needs to maintain the appearance of military dominance and the capacity to defend itself, as well as its allies and interests in order to preserve its hegemony. Conversely, if China can demonstrate a superior capacity in AWS and credibly undermine US military strength in the Pacific, it can discourage small-middle power states from bandwagoning against Chinese interests. Whether the United States retains the objective capacity to 'win' a future war against China is less important for neutral and allied states than the perceived power balance between the competitors (Fels, 2017). By undermining the United States as a security hegemon, China could encourage neutral states in the region to acquiesce to its regional expansion, to defect or even to encourage provocative self-help behaviours (Kraft, 2017).

Moving to a geo-economic perspective highlights that becoming a first mover in this space would give China or the United States greater influence over how AWS are perceived, deployed or potentially regulated once they begin to diffuse. This can be seen with remote piloted unmanned aircraft, where the United States did not sufficiently capitalise on its initial lead to secure dominance in the nascent export market, allowing China and Israel to assume leading positions, with greater influence over how early-majority adopters interacted with UCAVs. While becoming the first mover state does not guarantee dominance over the final innovation, there is a level of economic and political benefit to be gained, particularly from the perspective of maintaining regional influence in a hegemonic competition.

However, historically the fast follower adopter of an emergent disruptive innovation has proven to have an advantageous position. The fast follower runs less risk of pursuing a purported major military innovation that does not eventuate, can draw on operational, technological and integration lessons from the first mover and, in the case of AWS, could rely on sensors, software and concepts that were initially designed by the first mover, effectively shifting the significant burden of initial research and development during the incubation period. These advantages, as well as the known preference for emulation by lower capacity actors, accounts for the pattern of competing states being influenced by one another during the early deployments of a major military innovation, with the notable exception being those few cases where the developing state was able to maintain secrecy. Given the demonstrated Chinese track record of cyber-espionage, mandatory technology transfers and even blatant intellectual property theft, there is a clear incentive for the United States to limit what capabilities and systems it unveils, much less exports. However, developers must balance secrecy against the performative aspects of pursuing an offset strategy based on the emerging major military innovation. This is because, while an offset strategy requires that a developer reveal or hint at capabilities in hopes of deterring a would-be adversary, it must also maintain a sufficient hidden capability edge to acquire 'a war winning advantage if deterrence fails' (Work and Grant, 2019). This may prove particularly difficult in the case of artificial intelligence and autonomous systems because they are inherently more difficult to demonstrate to an adversary, especially in an escalating crisis situation. This is because the key enabler of AWS is its governing artificial intelligence software, meaning that the only way to objectively demonstrate capability to an adversary without actually deploying the AWS is to reveal internal coding, which states are unlikely to do given its comparative ease of diffusion and the risk that this would increase the system's vulnerability to cyber-attack or deterioration (Horowitz, 2019).

Examining recent Chinese engagement with ASEAN member states demonstrates that it has been actively competing for influence over potential coalition members in the region and encouraging uncertainty over the United States' continued role, itself a diplomatic aspect of China's wider preparation for competition with the United States. China's view of itself in the region reflects its traditional importance in Southeast Asia, which is viewed as its 'neighbourhood' or sphere of influence (Zhang, 2018). There is an interesting dual-nature to China's engagement with potential coalition partners, one that reflects the balance that must be struck between demonstrating capability and not being threatening to the point that ASEAN states would feel forced into direct balancing. At the core of this engagement is an approach that Zhang characterised as 'conditional reassurance', where conciliatory or economically beneficial diplomatic overtures are offered as incentives against a background of hard power deterrence. Whether China's preferred image as an essentially benign but powerful if provoked facilitator of regional growth has been undermined by its aggressive stance in the South China Sea has been debated (Zhang, 2018) but is clearly making its ASEAN neighbours wary.

8.3.3 ASEAN resistance to great power interference

Despite their comparative lack of power capacity, regional actors such as Indonesia or Singapore remain independent sovereign entities, whose traditional view has been that discouraging the establishment of monopolistic great power influence but remaining non-committed in the event of major confrontation is the best way to preserve their sovereignty. It is therefore unsurprising that one of the foundational objectives of ASEAN was to limit confrontation between great powers in Southeast Asia, a role that remains vital for maintaining regional security (Kraft, 2017). However, the centrality of ASEAN as a regional security actor has been challenged recently, with its shallow normative structure and consensus-based approach limiting the organisation's capacity to meaningfully contribute to contentious inter-state security issues. Therefore, it is more useful to view ASEAN and its structures as a forum through which leading ASEAN states can engage in indirect balancing in order to manage the dynamics of great power relationships with Southeast Asia, the use of intra-regional partnerships to coordinate responses on non-traditional regional security issues like transnational crime and climate change is an example of how these forums can be utilised to de-emphasise great power dynamics. While Indonesia and Singapore are certainly capable of independently exercising their power by attempting a more extensive adoption (whether or not it would be successful), their participation in the future hegemonic competition will depend on the dynamics of their relationships with both powers.

Rather than acting as a direct balancing alliance like NATO, ASEAN member states have demonstrated a preference for inclusive multilateral institutions (of which ASEAN is the key connector) that bring together actors in order to deny any one great power from asserting a dominant regional hegemony (Kuik, 2016). This is not typical balancing or bandwagoning behaviour, and ASEAN member states have been notably cautious to avoid the appearance of directly balancing either great power. Instead these forums are used to build cooperation and direct regional efforts to address non-traditional security issues. While an increasingly aggressive China has pushed individual Southeast Asian states to adopt closer defence ties with the United States, there is little chance this will be reflected in an ASEAN statement as it would make it more difficult for member states to hedge moving forward. For example, Singapore was chastised in 2016 following reports that its representatives were involved in a push to have the International Court of Arbitration decision mentioned in the joint statement from a Non-Aligned Movement meeting (Zhou, 2016). This incident further illustrated that, while it is unlikely that China would directly coerce an ASEAN member state to make a concession through military force, it has proven willing to leverage its economic advantage to deter ASEAN states from directly challenging its interests, even through a multilateral group.

In conclusion, despite the pre-eminent value ASEAN member states have placed upon non-interference and denying hegemonic dominance in the region, their reliance on the US security guarantee and Chinese economic partnership means that the organisation's role in limiting the impact of great power conflict must be viewed with a sceptical lens. Presently ASEAN states seem to be mirroring Indonesia and Singapore's preference for neutrality and continue to work towards limiting the potential for intra-regional conflict while denying dominance by either great power. While none of the ASEAN members are openly taking direct balancing action against China, the ongoing military modernisation efforts are reflective of their concern about the potential for regional conflict if economic competition builds into hegemonic conflict. ASEAN is not NATO or the EU, it lacks the necessary institutional rigour and structure, and the failure to issue a joint statement in 2012 was widely seen as indicative that China could successfully leverage individual members to scuttle collective action (Le Thu, 2019). Through the auspices of regional organisations like ASEAN, Singapore and Indonesia have greater potential to indirectly balance hegemonic powers.

8.3.4 Distinguishing a middle power approach to major military innovations

It is important to account for the fact that states, the neorealist 'billiard balls', differ significantly from one another in terms of resources, environment and goals (Finnemore and Goldstein, 2013). The preceding case studies, unsurprisingly, highlighted the stark disparity between the resource capacities of Singapore and Indonesia, and those of the United States or China. As an example, the US Department of Defense invested more in research and development of unmanned technologies (USD 9.6 billion) (Klein, 2018) in 2018 than Indonesia's entire defence budget that year (Institute of International Strategic Studies, 2019). Based purely on a comparative resource perspective it is understandable why the majority of the relevant literature has focused upon great powers; ASEAN states would simply not have the resources to quickly or effectively follow, much less lead, the United States in adopting the next aircraft carrier or nuclear missile.

There are three main problems with focusing on great powers in the incubation and post-demonstration point periods at the expense of minimising or dismissing the role of middle power states. The first is that this approach is predicated upon the false assumption that barriers (chiefly acquisitional, technological and operational) would bar regional middle powers from successfully adopting, integrating and deploying increasingly autonomous weapon systems. As argued earlier in this book, the adoption and emulation barriers at the entry-level are significantly lower in this case because much of the underlying technology is dual-use, and operation requires a lower skill floor than more advanced prior RMAs (such as carrier warfare).

Following from this, the second problem is that this approach neglects the fact that artificial intelligence, the core 'hardware' component underpinning the disruptive element of LAWS (their autonomy), is an enabling invention rather than a distinct self-contained platform, conceptually closer to the combustion engine than an aircraft carrier (Horowitz, 2018). It is therefore limiting to demarcate successful adoption solely in relation to whether a military can successfully integrate and deploy a LAWS in a direct combat role. Instead, middle power states can progressively integrate limited autonomous capabilities into their platforms over time as the underlying technologies continue to mature, diffuse and normalise. In this case, states can even capitalise on the growing civilian market to fill operational gaps (which has already been seen from Israel, Australia and the United States). The argument that Southeast Asian states would take this gradual approach rather than attempt a more traditional adoption response is supported by the rapid diffusion and proliferation of the identified precursor innovation, remote-operated UCVs, which is detailed in Chapter 4.

Finally, middle power states operate from a different geopolitical perspective to the United States or China and would therefore prioritise different capabilities when determining how to respond to a demonstration point. While the core purposes of adopting a major military innovation remain to offset either the strength of a rival or an adopter's weakness, middle power states are more concerned with leveraging technological superiority as a way to ensure their security and maintain prestige. In effect their interpretation of the universal state goal of survival places priority on preserving their position in the regional balance of power, rather than attempting to gain hegemonic status. Furthermore, unlike their larger cousins, middle power states generally know their likely opponent in future conflicts and do not necessarily need to be able to win in a potential war against a hegemonic great power, merely to deter aggression by raising the costs and risks to an attacker. Therefore, middle power militaries focus their efforts on maintaining a credible deterrent capability within a flexible force posture that can maintain their security and interests as well as support broader regional stability.

It is therefore important to recognise that 'adoption' will look notably different for small-middle power actors within a given regional structure (in this case Southeast Asia) than from the perspective of the great power states competing for dominance in that structure. Even successful adoption will likely be partial; beyond resource constraints there is little incentive for Southeast Asian states to attempt to fully emulate the capabilities being pursued by great powers over less advanced platforms or individual capabilities that still meet their less intensive operational requirements. For example consider the BAE Taranis UCAV; while its intercontinental strike capability could suit the requirements of the United Kingdom, it would be less likely to be adopted by Indonesia for the simple reason that the Taranis entails a highly resource-intensive acquisition process without offering comparatively more effective performance than cheaper, lower-capability platforms against the security issues prioritised by the TNI (internal security, intra-regional deterrence and maintaining regional stability).

Furthermore, a pure reliance on attempting adoption, even in a limited manner, is unlikely to be successful for regional middle powers; instead the pursuit of the disruptive military innovation must be integrated into a broader alliance-based diplomatic strategy. Horowitz suggested that the character of this response is determined by the pre-existing relationship between the middle power and the first mover (Horowitz, 2006). In the case of Indonesia and Singapore, their proclaimed preference for neutrality and current hedging behaviour indicates that they would prefer to continue to balance their linkages to the United States and China.

This section has argued that the regional middle powers would react differently to great power states to the emergence of disruptive military innovation, even if they are attempting a level of adoption. It is important to understand how middle powers will incorporate limited adoption of increasingly autonomous systems into a broader, externally focused response to the future demonstration point of LAWS. The following evaluation must therefore consider not only their adoption capacity but also their geopolitical context, great power entanglements and regional power relationships. Below is presented a list of capabilities that middle power actors in the Southeast Asia/Oceania region have publicly demonstrated an interest in developing (Table 8.1). Similarly, to the list of RMAs presented in Chapter 2, this list is not intended to be exclusive and is offered as an explanatory tool based upon the research conducted while preparing the preceding case studies. Basing the response evaluation upon the current capacities of Singapore and Indonesia provides an important grounding in the current technology and capabilities that will in turn enable a more robust analysis of state response and the post-demonstration point impact of LAWS.

8.3.5 Levelling effect of autonomous weapon system proliferation

Notwithstanding their rhetoric of neutrality and non-interference, the response of ASEAN member states to prior major structural power shifts is illustrative of the overriding, pragmatic objective of state survival, which has prompted ASEAN states to either bandwagon with a great power patron in the United States or form a balancing alliance among themselves. This reflects the structural neorealist view that smaller and middle power states simply would not have had the resources to engage in an arms race towards an emerging military innovation; for example there was no way for Indonesia to effectively compete in the post-Second World War period power projection paradigm. Even if it had overcome the material and organisational barriers to adopt aircraft carriers or nuclear weapons, the TNI would lack the scale and resources to effectively compete. However, in the case of autonomous weapon systems the diffusion barriers are sufficiently low that Indonesia and Singapore would be able to undertake limited adoption, and it is this democratisation of an emerging disruptive military innovation that is the crucial factor that makes autonomous platforms the next great leveller in international relations.

Unlike its historical predecessors the enabling element of autonomous military platforms is readily accessible software paired with dual-use sensors;[1] the complicated and expensive hardware components are merely secondary. Indeed, the physical weaponry carried by autonomous weapon systems is generally borrowed from comparable manned platforms. For example, the Apache attack helicopter fires significantly more Hellfire missiles annually than are used by armed UAVs, and the South Korean Super-Aegis II is equipped with a standard 12.7 mm machine gun. This book does not dispute the contention that the United States is the only state with the extensive informational infrastructure to support major, strategic-level campaigns using unmanned systems, although China and Russia are slowly building similar capacity. This book does, however, dispute the notion that a state requires extensive information infrastructure or a complex domestic arms production capacity to effectively deploy unmanned systems.

Whereas a state can be blocked from developing or acquiring physically advanced military platforms, increasingly autonomous weapon platforms represent

Table 8.1 Increasingly Autonomous Weapon Systems Being Pursued by States Active in Southeast Asia

Domain	Capability	Example of Platform or Pursuing Military	Capability	Example of Platform or Pursuing Military
Ground	Artificial intelligence enabled battlefield assistant	Australian Defence Force	Border surveillance/protection	Supervised sentry gun
	Strategic logistics	Hunter Wolf UGV	Fire support	Jaeger UGV
	Tactical logistics and casualty evacuation	MUTT	Tactical surveillance	Wasp III
Marine	Modular 'motherships'	Multi-Role Combat Vessel	Logistics	Venus-16 USV
	Harbour defence and force protection	USV Protector	Long-term surveillance	Wave Glider
	Strike capability	Sea Hunter	Protection or denial of SLOC	CAPTOR
Aerial	Tactical surveillance	Wulung UAV	Autonomous maintenance and repair stations	Singapore Air Force
	Medium–long range ISR	MQ-4C Triton	Tactical fire support 'Loyal Wingman'	DefendTex Tempest
	Semi-autonomous robotic pilots	PIBOT	Medium range strike and assessment	Royal Australian Air Force
	Identification and assessment of damage to military installations	Republic of Singapore Air Force		MQ-9 Reaper

a unique opportunity for developing states to compete with advanced militaries. This is because software diffuses much more rapidly than hardware, with its low transmission cost and comparatively low knowledge barrier. Furthermore, much of this software is not even military-focused and uses data from dual-use sensor technologies, such as LIDAR. Its inherently dual-use nature is reflected in the variety of non-state actors involved in its development (from researchers to Silicon Valley start-ups). This makes it much easier for states to acquire some form of autonomous weapon systems than previous major military innovations.

As an example, consider the difference between North Korea's Inter-Continental Ballistic Missile programme and their nascent UAV programme. Whereas the international community can use sanctions to restrict the transfer of missile components or high-grade fuel (aviation, jet or rocket), North Korea has already demonstrated a cyber-espionage capability that could be used to steal autonomous operation software from hundreds of civilian companies, many of whom are not even defence contractors. There is very little that the international community can do to prevent North Korea, or another rogue actor, from taking software from a commercial UAV or self-driving car and applying it to an armed platform.

Having access to even a limited array of unmanned platforms or autonomously operated weapon systems could allow Indonesia and Singapore to exert a much greater agency in the emerging hegemonic conflict between China and the United States, particularly if the acquisition was targeted towards specific disruptive capabilities to offset their resource weakness, which based on public ministerial speeches, Singapore has already realised. Again, ASEAN states do not have to be able to win or even effectively fight a war against China or the United States; rather they just need to be able to use this innovation, in combination with their strategic geographical locations to present a credible threat of harming a hegemonic challenger's capacity to compete. For example, while the Singaporean navy could not directly fight its Chinese counterpart, there are a number of ways the asymmetric deployment of unmanned or autonomous assets could enable the SAF to credibly threaten to cut China's maritime economic 'belt'. These range from the direct use of force through swarms of cheap, armed unmanned submarines or autonomous sea mines that only engage Chinese naval vessels; through to more indirect balancing options such as delaying customs approval, conducting offensive cyber operations, refusing to protect Chinese flagged shipping from non-state actors or even providing the United States with data on Chinese naval movements in the region based on a network of unmanned systems coordinated by an AI-enabled assistant.

The first major impact of autonomous weapon system proliferation from a hegemonic competition perspective will therefore be the levelling effect of unmanned platforms, giving smaller states and even non-state actors greater capacity to compete asymmetrically with larger states. The diffusion of unmanned systems will in turn lower the attractiveness of the US security guarantee, which is arguably the main inducement to ASEAN states to support their continued primacy in the region over China, especially if domestic production proves suffice for the more limited requirements of smaller states (as in the case of Singapore for example) or if a another early adopter offers comparable capabilities in cheaper platforms. Indeed, one of the vulnerabilities of the US Third Offset Strategy is that they have staked their capacity to maintain a valuable military offset on an inherently diffusive technology; they thus run the risk of heavily investing in an innovation that can be later matched by competitors at a far lower cost as fast-followers (Asaro, 2019). By enabling smaller states to disrupt the traditional power projection dominance of larger or more advanced states in this manner, AWS are quite unique as a major military innovation. While it is most likely that Singapore and Indonesia will continue to attempt to balance the increasing hegemonic competition in the Asia Pacific, the diffusion of increasingly autonomous unmanned systems and artificial intelligence software will give them greater freedom to manoeuvre during great power competition despite the increasing assertive positions of both China and the United States.

8.3.6 Increased ASEAN state agency and the risk of unexpected hegemonic conflict

Without slipping into the realm of technological determinism, the emergence of even the most disruptive major military innovations would not fundamentally alter the bedrock of international relations, which is the centrality of the comparative power generation capacities of its participants. Rather the disruptive effect of RMAs on prior hegemonic transitions and conflicts has been largely limited to the major actors involved in those transitions. Even at the regional level, major shifts in structural power do not always result in conflict among the middle powers, whose interest lays in protecting their relative position, resources and prestige rather than maximising their influence. This meant that when the Thucydides Trap led to conflict, middle powers operated as part of grand alliances secure at the local level barring proxy conflict, internal violence or intra-regional confrontation with an opposing coalition member. This limited uncertainty and thus the security dilemma of states like Singapore, Australia or Indonesia. However, the lower diffusion barriers of AWS mean that a number of state and non-state actors in this region could potentially gain access to an emerging and poorly understood (by policymakers) class of weapon system.

One of the significant risks associated with middle power states gaining access to increasingly autonomous unmanned systems or even derivatives is that this would inject another category of actors into the dangerous hegemonic transition period. Though China and the United States are increasingly clashing over influence, national interests and their trade relationship, policymakers in both states recognise that direct engagement between the superpowers would be severely damaging and are aware of the need to carefully manage provocative moves to prevent potential escalation (Allison, 2017). During the Cold War the practice of both the United States and the Soviet Union was to keep close control over their nuclear arsenals, even when they were forward-deployed or distributed to trusted allies. This limited the risk of a second-tier state unexpectedly provoking a nuclear conflict. Quite simply, the fact that only a small handful of actors had access to the nuclear weapons allowed for policymakers to rely on game theory to a greater extent than would be possible otherwise, and for commanders to partake in brinkmanship based on their understanding of the opposing military.

The difference is that when multiple states have access to a weapon system for which there are sparse norms of use and response accepted by both sides, there is a much more significant risk of provocation or the escalatory use of force between states leading to unintentional or unexpected conflict. Furthermore, the combinations of actors involved in this risk become far more complex than the essentially bilateral competition envisaged by that Hegemonic Transition Theory. Provocation or confrontation with autonomous weapon systems in Southeast Asia could occur between smaller states (on the same or opposing coalitions), or between a coalition member (such as Taiwan or Japan) and a hegemonic competitor (China). Without a common understanding of the 'correct' response, provocative acts (such as using a UAV to intrude in another state's territory or deploying unmanned surface vehicles to emphasise your claim to a particular set of disputed islands) could unexpectedly escalate into conflict.

This risk is further exacerbated by the tendency, discussed above, for fully autonomous systems to act unexpectedly and fail spectacularly, unlike a human operator there is no guarantee with current generation systems that an AWS would not trigger an escalation of force by engaging a target that, while legitimate, would not have been engaged by a human operator who infers that the action is performative rather than threatening. Examples would include when Russian fighter jets 'buzz' United States patrol aircraft with provocative but ultimately non-hostile intent (Lubold, 2018), or when South Korean aircraft fire over 300 shots close to the nose of a Russian electronic warfare aircraft intruding in their airspace as warning shots (Leone, 2019). Given the difficulty 'proving' the autonomous operation of a system to a potential adversary (Horowitz, 2019) an unintentional engagement in those instances, particularly between ASEAN member states, would be difficult to diffuse, especially if a human was killed in the incident.

Therefore, the second broader impact of autonomous weapon system diffusion in Southeast Asia is that it will necessitate a re-thinking of provocation and stability during hegemonic competition. While prior hegemonic transitions were focused upon the great power states that were competing for influence and used smaller states as supporters or fronts in proxy conflicts, the levelling effect of autonomous and unmanned platforms will give states like Indonesia and Singapore a greater capacity to deter great powers from overruling their regional interests. Yet this diffusion will also increase instability within the region as the traditional guarantor of security is challenged on a broader inter-regional level to defend its primacy in the Indo-Pacific.

It is important not to dismiss the potential for rising middle power states, especially in Southeast Asia, to capitalise on increasingly autonomous weapon platforms, particularly in cases where the United States could assist them to overcome the main barrier to large-scale strategic deployment in wartime, the need for increasingly sophisticated command and control capabilities. Given this potential, the discussion should turn to consider the most effective way to establish an international framework that will harness the potential of increasingly autonomous platforms towards increasing stability rather than exacerbating inter-state tensions.

8.4 Conclusion

In conclusion, the diffusion and proliferation of increasingly autonomous weapon systems in Southeast Asia will have significant, but not necessarily totally negative implications for the security of Singapore and Indonesia. This diffusion does however have the potential to influence the hegemonic competition between China and the United States within the region in such a way that would necessitate a re-thinking of the traditional view of hegemonic competition. This chapter explored the regional security and stability impacts of autonomous weapon proliferation in Southeast Asia, and the impact of this proliferation on competition between great powers for regional primacy.

The main purpose of this chapter was to demonstrate that this major military innovation is uniquely susceptible to diffusion and proliferation and that this prevents either China or the United States from guaranteeing that they would be able to maintain a sufficient technological superiority in the shifting power projection paradigm to impose their influence over regional powers during the transition period. This will incite a shift in the nature of future hegemonic competition away from a bipolar contest between great powers supported by coalitions of global-middle and regional-great/middle powers to a multipolar competition space where the dominant regional actors are able to exercise a greater level of agency and more practically stay neutral between the hegemonic camps.

Admittedly, it is not possible to completely eliminate uncertainty as to the exact features and limitations of lethal autonomous weapon systems at their demonstration point simply because we currently remain in the incubation period (Horowitz, 2006), however, by basing its analysis on the publicly known state of technology and regional geopolitics, this chapter has provided an effective outline of how the proliferation of increasingly autonomous unmanned platforms and derivatives would affect the balance of power and regional hegemonic competition in Southeast Asia that can be used as a resource to guide policymaking prior to a future demonstration point.

Note

1 This term is utilised with its technological, rather than innovation studies, definition. However, initially emerging operational praxes for the use of unmanned platforms indicate a significantly lower organisational capacity requirement than, for example, aircraft carriers.

9 Proposing a regional 'soft' normative framework for the safer deployment of AI-enabled autonomous weapon systems in Southeast Asia

9.1 Introduction

A major factor underpinning the concern around the regional security impact of increasingly autonomous weapon systems is the lack of established international law or norms governing the deployment of unmanned weapon systems. For example, a state could decide to send a strong, coercive diplomatic message to a neighbour by destroying or capturing an unmanned platform with the assumption that this would not necessarily spark the level of escalatory response that would result from destroying a manned vessel. Without established international law, behavioural norms or even a common definition of 'autonomous weapon system', capturing or destroying that unmanned platform could unexpectedly prompt an escalatory response. Furthermore, Southeast Asian middle power states are challenged with balancing the potential benefits of AWS to their security and stability against the risk of unmanned platforms proliferating into the hands of rival states or violent non-state actors.

This chapter focuses on the potential methods by which the diffusion of increasingly autonomous military technology could be regulated while still empowering states in this region. This chapter will demonstrate that the international community should turn its focus towards normative responses and consensus-building in order to create a genuine multilateral basis for future regulation. In the absence of an effective multilateral response to date, this chapter concludes by proposing the case for ASEAN member states to independently pursue a 'soft' normative framework that builds on the stalled CCW process to develop consensus-based regulatory measures that are derived from a functional assessment of autonomy.

When considering the impact of increasingly autonomous weapon systems within Southeast Asia and the potential to generate a limiting normative framework, the most relevant multilateral grouping is the Association of Southeast Asian Nations, a unique coalition of comparatively weak but fiercely independent states that holds regional security as one its key objectives. Established during the Cold War, one of ASEAN's foundational purposes was to maintain this independence and prevent further great power interference (Tang, 2018). As ASEAN developed, it was increasingly described as an important forum for coordinating security response among Southeast Asian member states.

DOI: 10.4324/9781003172987-9

ASEAN's diplomatic methodology (the 'ASEAN Way') reflects the nature of its membership, being both consultative and informal. This approach prizes a consultation and consensus-based approach that draws on cultural norms. Negotiations are usually conducted quietly outside of the public eye and favour non-interventionist responses (Goh, 2003). This approach has, however, been criticised as being too slow and cumbersome, especially in relation to potential flashpoints, such as the South China Sea. However, neutrality is not indicative of inaction or lack of engagement, and outside of traditional power confrontation (Tang, 2018), the nascent ASEAN security community has achieved some notable successes. Of particular interest to readers of this book is that ASEAN has focused on improving intra- and inter-regional multilateral military exercises, technology sharing and direct defence diplomacy.

9.2 Traditional measures for generating a framework for limiting impact of disruptive military innovation proliferation

There are two broad theoretical approaches for generating a framework for limiting the initial impact of major military proliferation. Firstly, the framework could be dictated and enforced by powerful states that gain a dominant early lead in the possession and development of autonomous weapon systems, albeit influenced by the persisting balance of power. However, as evidenced by previous revolutionary advances in military technology, including nuclear weapons, as the technology diffuses, the ability for the first mover or the dominant hegemonic power to control its use by other states diminishes. This effect is illustrative of the argument that 'bad policy by a large nation ripples throughout the system', and that the chief cause of structural power shifts is generally 'not the failure of weak states, but the policy failure of strong states' (Finnemore and Goldstein, 2013).

This effect was also evident in the case of unmanned aerial vehicles. The United States enjoyed a sufficient comparative advantage in the early 2000s that it could have theoretically implemented a favourable normative framework and secured itself a dominant export market position. However, as described above, it failed to do so until 2015 and 2016, by which time diffusion and proliferation were already occurring, driven by both other states and the civilian market. While the United States maintained a significant technological advantage at that point, it was no longer sufficiently dominant in the production of UAVs to impose its will on the market and China's rise in the Asia-Pacific was well underway. As a result, efforts by the United States to impose norms on the use of unmanned systems in 2015 and 2016 were only partially successful and had the unintended consequence of increasing the normative influence of China and Israel, who had assumed market dominance in the interim period.

In the absence of hegemonic leadership imposing a normative framework, we must turn attention to the international community. Supported by neo-liberal institutionalist theory, the second potential source for norm generation would be a multi-national institution (for example the United Nations)–led approach that aims to integrate controls under international humanitarian law. This approach recognises the increasingly interlinked nature of the global community from an economic and security standpoint. This process started for autonomous weapon systems in 2014 with an informal meeting of experts, followed by more formal proceedings at the Convention on Certain Conventional Weapons. Neither Indonesia nor Singapore is a direct participant in these negotiations (as non-signatories to the CCW), participating instead through the Non-Aligned Movement, which issued a statement that was interpreted as supportive of the regulation (not necessarily a ban). In the absence of significant progress towards a common understanding of how to meaningfully regulate autonomous weapon systems, with or without a developmental ban, this avenue towards an international normative framework does not appear promising.

Accepting that developing accepted international law to govern the deployment of increasingly autonomous unmanned platforms is unlikely to occur in the near future and that neither the development of autonomous technology nor the proliferation of unmanned platforms is likely to cease during the process of pressuring the international community into action, the third approach would be for regional organisations and security communities to take a leading role in developing norms and common understanding around the deployment of unmanned systems.

9.3 Potential forums for developing a normative LAWS framework and building regional resilience to post-demonstration point security shock

There are four ASEAN-led forums that could be utilised to formulate a regional normative framework for governing the use of autonomous weapon systems. These forums are the East Asia Summit, the ASEAN Regional Forum, the ASEAN Defence Ministers' Meeting and the ADMM-Plus. These forums have the capacity to build on the stalled work of the Convention on Certain Conventional Weapons' Group of Governmental Experts.

The first and the least suitable of these forums would be the East Asia Summit, a strategic dialogue forum with a security and economic focus, which was established in 2005. EAS brings together high-level state representatives in a diplomatic environment that encourages private negotiation and informal cooperation. The dual purposes of the East Asia Summit were to draw major powers into the Southeast Asian security environment and to create a platform for ASEAN member states to maintain influence with those powers.

To this end, the membership of the EAS extends beyond the ten ASEAN member states to include Australia, China, Japan, India, New Zealand, the Republic of Korea, Russia and the United States. These states are the primary actors in the region, representing a combined total of around 55% of the global population and GDP (Australia Department of Foreign Affairs and Trade, 2016). Furthermore, five of these states are known to be developing increasingly autonomous weapon systems. As part of their induction, all members were required to have signed the *Treaty of Amity and Cooperation in Southeast Asia*, a multilateral peace treaty that prioritises state sovereignty and the principle of non-interference, while renouncing the threat of violence (Goh, 2003). However, its broad membership means that this forum would suffer from similar barriers to a consensus as encountered in the UN-sponsored process. The inclusion of the United States, Russia and China would negate any advantage that could be gained from shifting to a regional focus. Finally, the EAS was not designed with the same defence focus as the following forums. Instead, the EAS is built around leader-to-leader connections and the summit itself, leading to an inability to facilitate concrete multilateral defence cooperation (Bisley, 2017).

The second forum to consider is the ASEAN Regional Forum (ARF), the first multilateral Southeast Asian security organisation (Tang, 2016). The ARF emerged in a post-Cold War environment, well before China had been widely recognised as a rising hegemonic competitor (Ba, 2017). The ARF was intended to be an all-inclusive security community, promoting discussion, peaceful conflict resolution and preventative diplomacy. While it has been used to promote regional efforts to reduce the illegal trade in small arms, the organisation's non-interventionalist security focus and lack of institutional structure limit its utility as a forum for developing a normative LAWS framework.

The ARF lacks the capacity to facilitate effective discussions towards a regional LAWS normative framework and has proven incapable to develop concrete responses to traditional security threats in the region, leading to frustration among its extra-regional participants. Ironically, the external membership of the ARF, currently 27 members, has been the main factor in frustrating these efforts. While the ARF's inclusive approach was a noble (and politically expedient) sentiment, it has naturally steered the discussion away from issues that would be sensitive to its members, contributing to its reputation as a 'talk shop' (Tang, 2016). Though the ARF has proven a useful tool for improving cooperation on non-traditional security issues and humanitarian aid, the participation of the United States and China has limited its capacity to meaningfully engage with major geopolitical flashpoints and has exposed divisions within the ASEAN membership (Kwok Song Lee, 2015). Therefore, while the ARF has played an important role in shaping the regional security architecture, it would be unsuitable for developing a regional response to LAWS.

The third mechanism through which a normative framework could be developed is the ASEAN Defence Ministers' Meeting. The establishment of the ADMM, and the complementary ADMM-Plus, was part of an institutional shift away from a diplomatic focus towards a functional one within the ASEAN Political Security Community (Tang, 2016). These forums were established as part of an Indonesian-led effort to maintain ASEAN centrality in the face of alternative security communities being mooted by external partners that were frustrated with the ARF (chiefly Australia and the United States). The ADMM directly links senior military leadership, intelligence services and security policy experts from each of the ten ASEAN member states through regular formal meetings that then feed into the Expert Working Groups of the ADMM-Plus (Ba, 2017).

There are two main reasons that the ADMM would be the best regional security forum through which to develop a normative framework that considers increasingly autonomous weapon systems. The first is that the ADMM is a comparatively neutral intra-regional institution that directly links the potential end-users of AWS within ASEAN without directly involving either China or the United States. The second benefit of the ADMM is that its core purpose centres on building trust and intensifying intra-regional military cooperation within the deliberately narrowed constraints of regional non-traditional security issues. Finally, as discussed below, the ADMM has already successfully developed and adopted advisory normative guidelines for the interaction of aerial and naval forces on the high seas that incorporate mutual definitions, procedures and practices to lower the risk of unintentional conflict or escalation in these domains, which a LAWS framework could be built around.

The final relevant forum is the ASEAN Defence Ministers' Meeting Plus, which is an extended, complementary version of the ADMM that incorporates the security services of eight extra-regional partner-states but remains officially ASEAN-centred.[1] The ADMM-Plus is a multilateral orientated grouping that is focused on practical defence collaboration in six key areas, each of which has an Expert Working Group. These areas of collaboration are maritime security, counterterrorism, military medicine, removal of mines, humanitarian and disaster relief and peacekeeping operations. Reviewing these focus areas highlights how ASEAN member states deliberately steered deliberations away from traditional security issues, reflecting the same geopolitical reality as in the ARF and EAS. However, the ADMM-Plus distinguishes itself with its role as a security-focused setting for defence policymakers to build trust, interoperability and relationships (Tang, 2016). Beyond policymaking, the ADMM-Plus facilitates valuable rotating collaborations between ASEAN and partner militaries to build trust directly between defence personnel, which would be necessary for any LAWS normative framework to succeed. As with the ADMM, this forum has the benefit of a more defined institutional structure that is built around Expert Working Groups (EWG) in each of these areas. However, while the EWGs are co-chaired by an ASEAN member state and an external participant on a rotating basis, the broader membership of the ADMM-Plus (particularly the United States and China) presents a greater risk of interference or delay in developing a normative framework than the more limited membership of the ADMM.

Overall, developing a normative framework for the safe deployment of autonomous and remote-operated weapon systems in Southeast Asia would be most likely to succeed if it was developed through a specifically established Expert Working Group within the ADMM forum. This would not be unprecedented, as the ADMM recently agreed to establish an Expert Working Group for cybersecurity. Unlike international law, there is no need for a region-specific normative framework to be formalised or publicly defended by participating states; nor would it need to be prescriptive or imposed on external actors. In this case the fact that the ADMM is not a traditional security alliance would not diminish the chances of this success because the region would benefit significantly from even a shared definition of autonomous weapon systems and a common normative framework for the acceptable use and appropriate responses to unmanned platforms. The ADMM actually already performs a similar trust-building and stabilising role within the region by facilitating direct defence diplomacy and multilateral training among the disparate Southeast Asian militaries and those of their external neighbours (Tang, 2016).

9.4 Analysing recent ADMM guidelines as a model

The ASEAN Defence Ministers' Meeting recently adopted two sets of relevant guidelines for military interaction on the high seas that provide valuable examples upon which a LAWS normative framework could be modelled. The *Guidelines on Air Military Encounters* (2017) (reproduced as Appendix A) were based on a concept paper written during the Philippines chairmanship, and the final document was published at the 12th ADMM the following year (while Singapore held the chair). This was followed by the *ADMM Guidelines for Maritime Interaction* (reproduced as Appendix B), adopted in July 2019.

There are three important aspects of these guidelines that are worth considering when pondering potential ADMM *Guidelines for the Deployment of Unmanned or Autonomously Operating Platforms*. The first is that both documents repeatedly and specifically note that their contents are 'non-binding and voluntary' and do not create any additional obligation under international law. Instead these guidelines are intended to reduce the risk of accidental or unintentional military escalation by establishing mutually agreed definitions and procedures that can be followed by member-state militaries and building mutual confidence between those militaries. Second, these guidelines make sensible use of existing international law and treaties as building blocks, deriving definitions, procedures and even technical specifications from previously established sources that are widely utilised (such as the United Nations *Convention on the Law of the Sea* [UNCLOS] or the *International Regulations for Preventing Collisions at Sea* [COLREG]) rather than 're-inventing the wheel'. Finally, neither document applies to the territory of member states (a clear concession to sovereignty concerns). Instead these guidelines apply solely to military interactions in the high seas, which complicates its application given the ongoing territorial disputes in the broader region. Importantly though, this concession highlights the fact that any framework on the use of LAWS would be unlikely to be successfully adopted if it was perceived to infringe on sovereignty without a commensurate benefit.

This final aspect could be overcome by the inclusion of a technology-sharing regime alongside the normative framework to offset the sovereignty concessions. While less appealing for Singapore technology transfer, access or even personnel exchange would be an influential offer to Indonesia, Vietnam or Malaysia. Further, as explored in the case studies, both Indonesia and Singapore are making a concerted effort to further develop their domestic military production capability but have identified areas where pooling resources would be valuable, while ASEAN already facilitates broader cooperation between the defence industries of its member states. It is also worth considering that the exchange of technology and personnel, as well as multilateral exercises, is the most common and effective method used to build interoperability and mutual trust among militaries, which would be vital for the safe deployment of LAWS.

Unfortunately, these guidelines were extremely short for multilateral policy documents. The *ADMM Guidelines for Maritime Interaction* is six pages long, while the *Guidelines for Air Military Encounters* is only seven pages in length. While the lack of detail in some points was discouraging, overall these guidelines still present concrete definitions and guidance on procedures. Given the comparative progress of underlying technology and the United Nations' discussions, even this level of agreement would be a significant step forward for the continued stability of Southeast Asia.

9.5 Conclusion

Generating a common understanding and increasing cooperation between states around unmanned platforms would reduce the short-term risk of escalation while the international community negotiates towards a more complete framework. This could remain a passive normative guidance framework (like the *ADMM Guidelines for Maritime Interaction*), or it could take a more proactive approach centred on a multilateral information and coordination agency modelled on the Regional Cooperation Agreement on Combating Piracy and Armed Robbery against ships in Asia. Without meaningful progress towards a mechanism for limiting the diffusion of artificial intelligence-enabled autonomous weapon systems, or a normative framework for preventing unexpected escalation, there is an understandable level of concern in the academic, policy and ethics spheres.

Concern about the potential negative impacts of autonomous weapons, however justifiable, should not be solely relied upon to support a position that autonomous weapon systems have no compensatory beneficial potential and should be pre-emptively banned. From a practical perspective, such ban would no longer be effective, given that the core enabling technologies for autonomous weapon platforms are of dual use and being developed by dozens of state and non-state entities. Yet more than that, this development also presents an opportunity for Southeast Asia as a region to apply an emerging technology to some of its most enduring non-traditional security threats, such as poverty, state instability and transnational crime, in a manner that reduces both the risk to human life and the risk of inter-state escalation.

Note

1 Australia, China, India, Japan, New Zealand, Russia, South Korea and the United States.

10 Conclusion and directions for future research

> Well, at any rate, judging from this decision of yours, you seem to us to be quite unique in your ability to consider the future as something more certain than what is before your eyes, and to see uncertainties as realities simply because you would like them to be so.[1]

This book has provided a detailed exploration of the factors that would influence AWS proliferation in Southeast Asia and the likely responses of leading regional powers. In turn, this provided a basis for critically analysing how AWS diffusion and proliferation would impact relations of power in this geopolitically crucial region, with an explicit focus on Indonesia and Singapore.

The core contribution of this book is addressing the lack of substantive engagement in the extant literature with the question of how the proliferation of increasingly autonomous weapon systems and AI would shape relations of power between non-great power states. While the potential moral, legal and ethical implications of LAWS have sparked a flurry of interest among scholars, NGOs, defence planners and policymakers, there is a comparative lack of published scholarly literature that links middle-power Southeast Asian states, military diffusion and increasingly autonomous weapon systems.

This book opened with the hypothesis that the uniquely low diffusion barriers of increasingly autonomous weapon systems would enable a rapid proliferation of related military technology to rising middle power states, and that in the absence of an effective framework governing their deployment the adoption of increasingly autonomous weapon systems by regional actors will raise the security dilemma of their neighbours and destabilise the emerging Sino-American hegemonic conflict as rising ASEAN powers leverage AWS to secure their neutrality. This concluding chapter will provide a systematic summary of how the preceding chapters have engaged with this underlying hypothesis through each of the three aspects of this book's core research question.

10.1 Understanding the influence of remote-operated UCV proliferation in the region

The first stage of this book focuses on how ASEAN member states responded to the proliferation of remote-operated combat vehicles. As the precursor innovation

DOI: 10.4324/9781003172987-10

for autonomous weapon systems, understanding how Indonesian and Singaporean policymakers reacted to UCVs provides a level of insight into their approach to autonomous weapon systems. Military decision-makers had historically proven to have an understandable penchant for viewing technological or doctrinal innovations through the lens of their prior operational experience, particularly when that involved a similar precursor innovation. In the case of LAWS, the precursor innovation is UCVs, which are distinguished by the fact that their 'critical functions' remain under the control of a human operator, albeit remotely. This book argued that Indonesia and Singapore have become genuine adopters of remote-operated UCVs, albeit on a notably more limited scale than the United States.

Enabled by lower entry barriers and guided by the operational concepts developed by more advanced powers, regional state and non-state actors have successfully utilised remotely piloted UCVs for surveillance and even to apply force. The willingness of ASEAN member states to build on or purchase civilian platforms to fulfil similar operational functions to military-grade small UAVs indicates a capacity to adopt lower-capability platforms to fulfil perceived operational needs that have been largely overlooked by existing research on LAWS. This evaluation illustrated the process by which the resource barriers for entry-level adoption of increasingly autonomous weapons will fall as the enabling technologies mature.

This analysis also highlighted the key arms-transfer relationships, and organisational structures responsible for military experimentation, procurement and modernisation that will allow Singapore and Indonesia to access increasingly autonomous systems where they are unable to attain domestic production capacity. Given the compression of the incubation period and the level of intersection between remote-operated and autonomous systems, it is unlikely that these organisational structures will have radically shifted in the period between UCAV proliferation and the current AWS incubation period.

Beginning with the precursor, innovation is valuable for this book's analysis because it offers an insight into how senior officers in the TNI and SAF are likely to perceive increasingly autonomous systems and indicates where predetermined path dependencies would interfere with attempted adoption of increasingly autonomous weapon systems. Furthermore, this analysis supports an argument that the dual-use nature of the critical enabling hardware components of autonomous weapons (chiefly AI and various sensor types), combined with the lower operational requirements of regionally and internally focused ASEAN militaries, necessitates a re-evaluation of the criteria used to indicate 'successful' adoption when examining middle power states. Finally, this approach provided a framework for controlling the actor and observation variables in the subsequent examination of LAWS by using the same methodology and case studies (Singapore and Indonesia).

10.2 Comparative outline of Indonesian and Singaporean adoption capacity

The core of this book has been an evaluation of the adoption capacity of Indonesia and Singapore, leading ASEAN member states. The relevant variables were resource capacity and organisational capital capacity from adoption capacity theory as well as the additional variables of security threat environment (Schmid, 2018), demonstrated capacity to develop or emulate a specialised operational praxis and the receptiveness of the populace to autonomous systems. The results of these evaluations were then used to determine whether Singapore and Indonesia would be successful in adopting their preferred response option following a future LAWS demonstration point.

The first variable affecting Southeast Asian engagement with increasingly autonomous weapon systems was the regional security environment. The main security threats facing both Indonesia and Singapore include terrorism, piracy, organised crime, internal instability, regional military modernisation and an increasingly assertive Chinese military posture in the region. Strategic documents from both militaries confirm that their chief concern is internal and non-state actor threats. This focus would shift interest towards efficient surveillance platforms, tactical-level capabilities and artificial intelligence-assisted information interpretation. Finally, the SAF and TNI remain wary of each other. While this has historically discouraged overly aggressive acquisitions, it does suggest that the deployment of AWS by one state would spur the other to either emulate or counter.

From a pure resource capacity standpoint, the SAF and TNI are supported by the largest military budgets in a region characterised by significant recent military build-up. While dwarfed by larger, advanced middle power states in the broader Indo-Pacific (such as South Korea, Japan and Australia), this expenditure remains significant in the context of Southeast Asia. In terms of converting financial resources into military capability, Singapore is able to draw on the most sophisticated defence production capability in Southeast Asia, while Indonesia's domestic arms production industry has been undermined by consistent underinvestment. Finally, both states maintain relevant foreign arms acquisition pipelines, while the rising capability of civilian equipment reduces their reliance on traditional military development.

Of the two case studies, Singapore has the superior organisational capital for supporting the limited adoption of increasingly autonomous systems. The recent investment in its defence technology community has promoted relevant technological development, and the Future Systems and Technology Directorate is challenging the structural preference for evolutionary innovation, which has historically been the principal barrier to the SAF pursuing revolutionary innovations. In contrast, the pursuit of increasingly autonomous systems is not reflected in the critical task focus of the senior leadership of the Indonesian Army, which has an unusual level of influence over its civilian overseers. Furthermore, chronic underinvestment has limited the capacity of the Indonesian innovation base to develop autonomous systems. The comparative lack of organisational capital capacity limits the adoption potential of Indonesia compared to Singapore.

Thirdly, despite the dearth of published public opinion data, there is evidence to suggest that the adoption of autonomous weapon systems would not be actively opposed by the domestic population. Firstly, both governments have passed regulations allowing for state use of remote-operated platforms in various security and civilian roles without notable public opposition. Furthermore, commercial entities have begun to incorporate autonomous systems; Indonesian businesses have the highest uptake of artificial intelligence in Southeast Asia, while Nanyang Technological University is scheduled to begin trials of automated (but initially supervised) buses in late 2019 (Wei, 2019). Thirdly, both states have remained at a distance from the ongoing CCW meetings, content to participate through the auspices of the Non-Aligned Movement; indeed, both states remain non-signatories to the convention. This indicates a lack of domestic pressure on Indonesia and Singapore to be seen to be contributing to the regulation of LAWS.

It is also worth noting that both Indonesia and Singapore have demonstrated a capacity to make military decisions divorced from effective public scrutiny, both in terms of expenditure and deployment. The TNI remains an influential force in domestic politics, bolstered by its ongoing practice of deploying military units across the archipelago alongside civil authorities, which was reflected in the strong performance of a retired general in the 2019 presidential elections (Yulisman and Salleh, 2019). While the SAF operates under a very different civil–military relationship, the state's security focus and stringent controls over the populace limits effective public opposition to military policy. As a result, the ruling political party has been able to consistently allocate significant resources to military modernisation even in periods of extended interstate peace and slowing economic growth. There is little evidence to suggest that there would be a significant departure from this norm in the case of autonomous weapons and the Next-Gen SAF.

Finally, while both states have demonstrated a capacity to emulate the operational praxes of more advanced states, the SAF has the distinct advantage of experience drawing on foreign strategic concepts as part of all three of its prior generational evolutions. The SAF also possesses a demonstratively more advanced doctrinal development and training capacity. By comparison, Indonesia's demonstrated emulation experience primarily relates to absorbing production capabilities from platforms initially acquired from foreign states. There is less evidence of meaningful engagement on doctrinal development, and the TNI remains heavily influenced by senior and recently retired army leadership, although there have been suggestions that the modern generation of post-Suharto officers are markedly more open to pursuing revolutionary operational concepts.

Overall, this book asserted that Singapore has a markedly greater capacity to attempt limited adoption than Indonesia. While ongoing military modernisation efforts by both states have provided sufficient military expenditure, the chronic commitment-investment gap in Indonesian defence spending and its ongoing underinvestment in the air force, navy and military research and development has meant that TNI investment in military modernisation remains poorly targeted, army focused and inefficient. Furthermore, although modernisation is slowly occurring, Indonesia's arms industry remains plagued by corruption, underinvestment and limited in sophistication. Indonesia could not indigenously produce advanced autonomous weapon platforms that would meet the definition utilised by this book, and would, therefore, have to rely on derivatives, foreign arms purchasing arrangements or technology transfer to attempt even a limited scale adoption of AWS. Equally importantly, this evaluation indicated that the modernisation efforts required to capitalise on the value of increasingly autonomous systems for regional security do not align with the critical task focus of the army-dominated TNI senior organisational structure. Given the continued influence of the TNI on domestic politics and power relations, it is unlikely that the civilian leadership would be able to direct the TNI to shift its focus to pursue increasingly autonomous naval or aerial platforms as part of the Global Maritime Fulcrum strategic concept.

The Singapore case study further supported the position that a distinct understanding of middle power offset would be required to account for ASEAN state adoption of increasingly autonomous platforms. While Singapore has identified the operational value of increasingly autonomous systems and can draw upon a significantly more advanced domestic arms production capacity and security innovation apparatus, its scale as a middle power means that it could only commit a fraction of the resources of great power states to pursuing autonomous weapon systems. Aside from this resource gap, however, Singapore is well placed to become a limited fast following adopter of increasingly autonomous unmanned systems, which feature prominently in the Next-Gen SAF strategic framework. Although the SAF's hierarchical preference for slow, carefully reviewed investment decision-making would theoretically limit its capacity to effectively innovate, the SAF has identified the importance of unmanned systems for maintaining their critical offset despite an impending demographic shift. Historically, the SAF has proven very capable of experimenting and emulating weapon systems that are identified as a priority in this manner .

10.3 Determining most effective response to future LAWS demonstration-point

While there are clear distinctions between the capacities of Indonesia and Singapore to take a proactive stance in response to the emergence of increasingly autonomous systems, the fact remains that neither state has the resource capacity to meaningfully compete against China, the United States or Russia to achieve first-mover status in the development and demonstration of a fully autonomous weapon system that would meet the definition used in this book. The main argument of this book, however, is that this inability is irrelevant and misleading in that the recognition that a middle or minor power could not meaningfully compete can lead into the assumption that their participation in this space is of little security impact.

This book has instead asserted that, once the definition of 'adoption' of a disruptive military innovation is shifted to account for the smaller scale and requirements of middle powers, it is clear that both states possess sufficient economic resources to attempt this limited adoption, challenging the implicit assumption in the existing literature that only the response of great power states is significant with LAWS.

Table 10.1 Indonesia and Singapore: Adoption Capacity Evaluation Results

State	Resource Capacity			Receptiveness of Domestic Audience
	Financial Capacity	*Domestic Military-Industrial Base*	*Foreign Arms Acquisition*	
Indonesia	Medium	Low	Medium	Anticipated Yes
Singapore	Medium	High	Medium	Anticipated Yes
	Organisational Capital Capacity			**Demonstrated Capacity to Develop or Emulate Specialised Operational Praxis**
	Critical Task Focus	**Level of Investment in Experimentation**	**Organisational Age**	
Indonesia	Internally focused, Army-dominated, non-state actors and internal conflict	Low	Little experience with interstate war, weak civilian influence in civil–military relationship.	Improving capacity to integrate platforms but little evidence of prior doctrinal emulation. Starting to emulate networked warfare.
Singapore	Regional security, internal security, projecting credible deterrence, maintaining technological offset	High	Hierarchical and heavily structured, three prior significant force structure evolutions	Strong history of emulating and adapting strategic concepts developed by other states, especially the United States.

Adapted from Horowitz (2010).

Nor, however, is adoption the only response available to ASEAN member states, particularly given the concurrent hegemonic competition unfolding during the incubation period of LAWS. Chapter 8 argued that limited adoption would be in the interests of both states as they attempt to maximise and preserve their relative status following a future demonstration point of LAWS, but also demonstrated that adoption would be most likely to succeed as the secondary component of their response. Given the history and existing balance of power in Southeast Asia, states like Singapore and Indonesia will be incentivised to pursue LAWS to secure their own neutrality between China and the United States, attempting to carefully balance commitments to both great powers while continuing to exert influence at the regional level in pursuit of recognition as an emerging great power (Indonesia) or survival and stability (Singapore). While not all ASEAN member states would have the capacity to effectively adopt platforms that meet this book's definition of LAWS, adoption would not necessarily be their optimal response, and should not preclude further analysis of how these states will nevertheless impact regional relations of power post-demonstration point.

10.4 Understanding the impact of increasingly autonomous weapon systems on relations of power in Southeast Asian security environment

The final research question guiding this book focused on understanding the extent of the impact of LAWS proliferation in Southeast Asia. Contained in Chapter 9, this section of the book examined the regional security impact of autonomous system proliferation to rising Southeast Asian middle powers on hegemonic competition and proposed a regional normative approach for limiting the risk currently inherent in deploying LAWS in Southeast Asia.

At the regional level, this book has demonstrated that, while the proliferation of increasingly autonomous systems has clearly negative and destabilising potential, AWS also offer policymakers significant potential benefits for responding to regional security risks. The major regional risk arising from autonomous technology diffusion is the potential to exacerbate existing tensions as ongoing regional military modernisation efforts are combined with an innovation that has been proclaimed by China, Russia, the United Kingdom and the United States as crucial to the future of warfare. Without an effective control framework or greater efforts to establish mutual trust among ASEAN members, the adoption of autonomous weapon systems (or declared intention to do so) by a Southeast Asian state would raise the security dilemma of its neighbours, creating a destabilising cycle of arms procurement if not a formal 'arms race' (Horowitz, 2019). Furthermore, the potential use of artificial intelligence-enabled agents to spread disinformation or conduct cyber operations, as demonstrated by Russia, would be particularly inflammatory in a region that emphasises non-interference and sovereignty. However, this book has also demonstrated that, if deployed safely, there are meaningful benefits to be gained from deploying increasingly autonomous platforms that deserve consideration.

It is becoming increasingly apparent that artificial intelligence-enabled autonomous weapon platforms are an emerging disruptive military innovation with the potential to shift the dominant paradigm for the use of force. This kind of shift has historically created an opening for a rising great power to challenge the regional hegemon for influence, resulting in tension, competition and usually conflict. While smaller states have historically been unable to pursue adoption of an emerging major military innovation, the lower entry barriers and higher diffusion speed of the core enabling technologies indicate the levelling potential of AWS proliferation.

Combining limited adoption into a primarily external response would increase the ability of Singapore and Indonesia to maintain their neutrality and independently pursue their regional interests to a greater extent than in prior cases of hegemonic transition, where middle power states were generally relegated to a subordinate position of coalition membership. While theoretically a positive result, this increased agency would increase the risk of conflict at the regional level as states feel threatened by their neighbours without the stabilising influence of a dominant hegemon. Furthermore, this would increase the risk of conflict between coalition members, which could, in turn, draw China and the United States into an unexpected conflict. Overall, this section demonstrated the potential for AWS proliferation to impact the emerging hegemonic competition between China and the United States in a way that challenges the traditional view of hegemonic power transition and further emphasises the disruptive impact of LAWS at the regional level.

10.5 Limitations and contributions

The core purpose of this book has been to critically engage with the disruptive impact of middle power states in this specific region, gaining access to increasingly autonomous weapon systems. Inherent in this purpose is four limitations imposed on the scope of analysis, which were explained in the introductory chapter. The first is the narrow geographic focus on Southeast Asia, examined through the lens of these two case-study states. The second limitation of the preceding analysis was beginning with the assumption that LAWS could be characterised as a disruptive military innovation (noting that some recent research has shifted focus to artificial intelligence, which in this book is characterised as a core 'hardware' component of LAWS). The third was that this book did not attempt to address the gap in published public opinion data regarding attitudes towards autonomous weapon systems among ASEAN states. Finally, this book has not engaged directly with non-state actors or the sub-national impacts of autonomous weapon system proliferation. While this research would be a valuable contribution to the field, space and resource constraints precluded its inclusion in sufficient depth; however, the author intends to pursue this research in future publications.

Within the scope of these analytical boundaries, the main contribution of this book to existing scholarly literature and the ongoing policy discussions on LAWS lies in its application of military innovation theory to middle power state engagement with increasingly autonomous weapon systems in a Southeast Asian context. Although a significant body of literature has emerged in recent years related to LAWS, the majority focuses questions around regulating AWS and their inherent morality. Where prior research has considered the impact of AWS on relations of power, it has generally done so with a focus on great power and active AWS-developer states. This book has instead asserted the importance of understanding the perspectives of non-great power states in this debate and accounting for the impact of their engagement with AWS, which would differ considerably from the current great-power-centric projections and analysis.

Taking this a step further, however, this book's focus on comparatively small, middle power states in an important but under-researched region reflects its main contribution to the existing body of scholarly work on the lifecycle and diffusion of military innovations. While this book admittedly takes the liberty of assuming that increasingly autonomous military technologies will eventually come to be recognised as a major military innovation, it is the focus on smaller state participation, its assertion of a distinct smaller-power version of offset strategy and adoption-response, and the introduction of precursor innovations that are significant departures from the existing military innovation literature.

While there is a well-documented, albeit still debated, linkage between major military innovations and great power competition, its study has largely neglected the role of middle power and minor states within great-power coalitions and minimised the role they play in the proliferation of innovations (be they military or commercial). This book has challenged the tendency to disregard the role of middle power states in the emergence of major military innovation and suggested an alternative perspective for analysing adoption within a more limited scale. This approach allows for a far more detailed analysis of the impact of major military innovation on transitions of hegemonic power than prior examinations that have typically assumed that the agency of such powers was subsumed by the goals of coalition-leading great powers.

This leads into a further contribution of this book to the scholarly literature, which has largely ignored the application of military innovation studies to states within Southeast Asia. While a small cohort of scholars (principally at the S. Rajaratnam School of International Studies) that more broadly researches the process of military modernisation and innovation within Indonesia and Singapore, there is comparatively little recent research that explicitly links Southeast Asian states, military diffusion theory and autonomous weapon systems.[2] To my knowledge, this is the first major scholarly work to apply military innovation theory to determining the impact of autonomous weapon system diffusion in this region.

This book has attempted to address these gaps in the scholarly literature, contributing to an increased understanding of how ASEAN member states are responding in the incubation period of increasingly autonomous weapon systems and the potential impact of their participation in the post-demonstration point diffusion of LAWS.

10.6 Directions for future research

This book has highlighted the potential for further research in three other areas, which largely stem from increasing its analytical scope. The first direction for future research would be to apply this approach to another geographic region or collection of non-great power states. While this book focused on Southeast Asian states, it is more broadly applicable to improving scholarly understanding of how major military innovations with low adoption barriers proliferate within a complex regional environment. For example, there would be value in applying this book's analytical framework to case-study states in Sub-Saharan Africa or East Asia. In the former case, an analysis focused on Uganda and Nigeria as leading states within the African Union (which has been expanding its multinational security function with mixed results) would provide valuable insights into how increasingly autonomous system adoption would influence African stability. Indeed, combining the findings in this book and those of a future African-focused study would provide valuable insight into the likely impact of autonomous systems proliferating to any number of the variety of non-state armed groups active in either region. In the latter case, an analysis centred on East Asia could incorporate the emerging security-focused Quad alliance and delve deeper into the potential role of larger middle powers such as Japan. Arguably of greater interest, however, would be that an East Asian-focused application of this book's analytical framework could shed much-needed light on the potential impact of autonomous weapon system proliferation on the Korean Peninsula, arguably the most important geopolitical flashpoint in today's security environment.

The second additional research avenue centres on determining public opinion towards increasingly autonomous military technologies outside of the United States. In addition to surveying the general public based on realistic scenarios for the deployment of autonomous systems in a military setting, it would be valuable to conduct interviews with ASEAN member state military personnel at both junior and senior levels. The final direction for future research would be to shift its analysis to focus upon non-state actors and subnational relations of force. Southeast Asia would be a suitable geographic focus for such research because there are numerous violent non-state actors active in the region and responding to non-traditional security threats features prominently in recent security documentation from ASEAN member states. Given the levelling effect of artificial intelligence-enabled weapon systems, there is a genuine risk of violent non-state actors capitalising on this innovation (or more likely a derivative) to offset the power advantage of regional militaries and security agencies. Indeed, limited adoption by terrorist groups, insurgencies, NGOs, transnational companies and organised criminal groups has already been seen in the case of remote-operated unmanned aerial vehicles.

10.7 Final overview

This book has brought together existing literature examining military innovation and diffusion, disruptive military innovation and international transitions of power, applying elements of each through a defensive neorealist lens to the proliferation of increasingly autonomous systems among middle power states in Southeast Asia. Its main purpose has been to explore and analyse the impact of middle power states incorporating limited adoption of an emerging major military innovation into their early post-demonstration point response on regional and hegemonic relations of power. As major powers lose their capacity to maintain a dominant edge over this innovation, there will be a shift away from bipolar hegemonic conflict towards a multipolar competition space where ASEAN members will have a greater capacity to continue their preferred balancing act.

While it is not possible to completely eliminate the uncertainty inherent in analysing the response to a future demonstration point, this book has minimised the impact of this uncertainty by basing its analysis on publicly available data and the current state of technology at the time of writing. As the demonstration point draws closer, the existing barriers to adopting or developing increasingly autonomous weapon systems will fall as the underlying technology matures and diffuses, while the regional geopolitical tensions and hegemonic competition are unlikely to fundamentally reverse their escalating trajectory in the remainder of the incubation period.

However understandable, concerns with the potential negative impacts of developing weapon systems with increasing independent control over their critical functions should not prevent deeper scholarly investigation into the potential impacts of proliferation in the event that an effective developmental ban does not come into effect. This book has engaged critically with the diffusion potential of increasingly autonomous and unmanned platforms into Southeast Asia from the perspective of the rising middle power states that will play a crucial role in the event that hegemonic competition between the United States and China erupts into conflict. More than that, however, this book has also highlighted the potential benefits of incorporating autonomous systems into the security apparatuses of these states, who are increasingly struggling with balancing their security, the perceived need to modernise militarily alongside their economic growth, and the need to protect against a wide range of violent non-state actors.

As increasingly autonomous weapon systems develop through the incubation period, it will become increasingly vital that the voices of the modern Melians be given greater weight in the international response. The levelling effect of this innovation challenges the traditional, neorealist view of power transition and the Thucydides Trap. Despite this risk, a pre-emptive ban under international humanitarian law assumes that there is no potential compensatory benefit from this innovation, and its practical viability faces serious challenges. Instead, we must focus on the capacity of middle power states to internally regulate their use of increasingly autonomous technology, control the proliferation to violent non-state actors and maintain the intraregional trust required to prevent unintentional conflict or provocation.

Notes

1 Thucydides, quoted in Freedman (2017).
2 The literature review for this book identified less than 20 distinct authors who published research directly relating to the military innovation process of Indonesia or Singapore between 2015 and 2019.

Appendix A: ASEAN defence ministers' meeting guidelines for air military encounters

GUIDELINES FOR AIR MILITARY ENCOUNTERS

Introduction

1. The rising growth, development and prosperity of countries in the Asia-Pacific have led to an increase in maritime and air traffic in the region. Specific to the air domain, the International Air Transport Association (IATA) estimates that commercial air traffic will double to 7.2 billion passengers in 2035, with more than 50% of this growth – or an additional 1.8 billion passengers – coming from the Asia-Pacific. With prosperity, regional countries are also modernising their militaries, including air forces, both for their own upgrading and to meet the demands arising from new regional security challenges. Looking ahead, defence expenditure in the Asia-Pacific is projected to rise by 23%, to more than US$530 billion in 2020. These trends will increase congestion in the air.

2. Since its establishment in 2006, the ADMM has made significant progress in promoting strategic dialogue and cooperation against common regional security challenges. Today, the ADMM cooperates in wide-ranging areas from HADR to crisis communications, and crossed a milestone last year when we commemorated the tenth anniversary of its establishment.

3. Recognising that the safety and security of air lanes are important for the growth and prosperity of countries, it is important to consider developing a set of guidelines that military aircraft can practise. These guidelines will help reduce the likelihood of encounters or incidents spiralling into conflict in the event of a miscalculation. Such guidelines would help reinforce the spirit of the ASEAN Political-Security Community Blueprint 2025, which calls on all ASEAN member states to promote shared values and norms as well as principles of international law, in building a rules-based community. Such guidelines will also adhere to the existing aviation standards promulgated by the Convention on International Civil Aviation (the Chicago Convention), the International Civil Aviation Organisation (ICAO) and the International Code of Signals (ICS) which all ADMM-Plus countries have subscribed to, and observe recognised international principles concerning military and

state aircraft governed by the 1982 United Nations Convention on the Law of the Sea (UNCLOS). Such guidelines will also complement the Code for Unplanned Encounters at Sea (CUES) adopted by the Western Pacific Naval Symposium, which naval aircraft of ADMM-Plus countries already observe.

4. This paper puts forth a broad set of principles for guidelines on air encounters between military aircraft, as well as operational guidelines.

Principle

5. The guidelines shall be non-binding and voluntary and serve as a practical confidence-building measure for the militaries to improve operational safety in the air.

6. The guidelines shall be applicable for unintentional encounters in flight between military aircraft over high seas, ensuring safe separation to avoid creating a safety hazard. To determine safe separation, military aircraft should comprehensively consider their own national rules and relevant international guidance.

7. The guidelines shall reaffirm the principles of Article 2 of the ASEAN Charter.

8. The guidelines shall respect the independence, sovereignty, territorial integrity of all states.

9. The guidelines shall be based on ASEAN principles of transparency and mutual trust, and shall be in accordance with relevant national laws, rules and regulations, and international laws.

10. The guidelines shall reaffirm states' commitment to resolve disputes through peaceful means without resorting to the threat or use of force in accordance with internationally/universally recognised principles of international law, including the United Nations Convention on the Law of the Sea.

11. The guidelines shall uphold all existing maritime and aviation arrangements between states, as well as between states and other organisations including, but not limited to, UNCLOS and CUES.

Adoption and review

12. The above framework consisting of principles for the air guidelines, as well as operational guidelines on abiding by existing aviation conventions and rules, safe and professional communications, standard flight procedures, encouraging mutual trust and confidence in the air, and contingencies and emergencies will be submitted for the ADMM's adoption through the ADSOM WG and ADSOM.

13. This set of guidelines is also available for implementation by non-ASEAN member states' military aircraft.

14. The framework as well as operational guidelines are evolving documents, and may be reviewed and revised with the consensus of the ADMM. Any

derivatives, or annexes, to operationalise the guidelines are to be negotiated at a later stage with thorough consideration to applicable situations.

Conclusion

15. As key stakeholders in the region, it is the responsibility of militaries among ASEAN member states to ensure the safe and smooth conduct of encounters between our military aircraft, particularly in light of increasing air traffic in the region. This will help to promote a safe, secure and peaceful operating environment in the region to allow the benefits of the global commons to be shared and enjoyed by all.

. . . . .

Annex A

Annex on observing existing aviation conventions and rules

1. Military aircraft[1] should, as necessary, operate consistent with existing and relevant aviation conventions and rules. This includes the Convention on International Civil Aviation and its Annexes, as well as UNCLOS. In particular, subject to international law, military aircraft are entitled to the rights and freedom of navigation, overflight and other internationally lawful uses of the sea related to those freedoms in high seas.

Annex on safe and professional communications

1. Military aircraft that encounter each other in flight should ensure navigation safety through professional airmanship, including the use of appropriate communications. The relevant references for communication and contact between military aircraft are the ICAO Annexes, ICS and the Radio Regulations of the International Telecommunications Union.
2. Military aircraft should communicate actively, including with the appropriate air traffic services units, in the interest of flight safety, through providing details such as identity and any other information related to flight safety should their aircraft be engaged in an activity that could affect the safety of nearby military aircraft.
3. Military aircraft shall establish two-way communication as necessary and in accordance with relevant international aviation rules and conventions.
4. Military aircrew should refrain from the use of uncivil language or unfriendly physical gestures.
5. Communications between military aircraft during an emergency may be conducted in accordance with the Convention on International Civil Aviation and its Annexes.

Annex on standard flight procedures

1. When military aircraft intentionally approach other military aircraft for the purpose of identification, interrogation, verification or escort, the pilots should operate with professional airmanship and exercise prudence for the safety of other approaching military aircraft. Meanwhile, each military aircraft should avoid reckless manoeuvres.
2. To determine safe separation, military aircraft should comprehensively consider relevant international guidance, and factors including the mission, meteorological considerations and flight situation.
3. Military aircraft should refrain from interfering with the activities of other states. However, military aircraft always enjoy the rights and freedom of navigation, overflight and other internationally lawful uses related to those freedoms in high seas.

Annex on encouraging mutual trust and confidence in the air

1. The aircraft commander of a military aircraft is responsible for determining whether his or her aircraft is threatened by another aircraft. That determination could be made through communicating actively with other military aircraft in the vicinity and with the appropriate air traffic services units that operate in the area.
2. Pilots should also consider the potential ramifications before engaging in actions that could be misinterpreted.
3. A prudent pilot should generally avoid: (a) actions that impinge upon the ability of other military aircraft to manoeuvre safely; (b) approaching other military aircraft at an uncontrolled closure rate that may endanger the safety of either aircraft; (c) the use of a laser in such a manner as to cause harm to personnel or damage to equipment onboard other military aircraft; (d) actions that interfere with the launch and recovery of other military aircraft; (e) aerobatics and simulated attacks in the vicinity of other military aircraft and (f) the discharge of signal rockets, weapons or other objects in the direction of other military aircraft encountered, except in cases of distress.

Note

1 Military aircraft include manned and unmanned fixed-wing aircraft, rotary-wing aircraft and helicopters of militaries.

Appendix B: ASEAN defence ministers' meeting guidelines for maritime interaction

ASEAN DEFENCE MINISTERS' MEETING (ADMM)

GUIDELINES FOR MARITIME INTERACTION

Background

1. Through a joint declaration, the ASEAN defence ministers agreed to 'undertake practical measures such as protocols of interaction and direct communication channels to reduce vulnerability to miscalculations and to avoid misunderstanding and undesirable incidents at sea' during the 9th ASEAN Defence Ministers' Meeting (ADMM) in 2015.
2. The ministers further agreed to 'practice and observe international protocols such as Code for Unplanned Encounters at Sea (CUES) and commence work on crafting protocols of interaction to maintain open communications to avoid misunderstanding and prevent undesirable incidents' as reflected in the 2016 ADMM Joint Declaration.
3. The ADMM then adopted the Concept Paper on Guidelines for Maritime Interaction on 23 October 2017 in Clark, Pampanga, to reduce vulnerability to miscalculations and avoid misunderstanding and undesirable incidents at sea.

Objectives

4. In line with the Concept Paper on Guidelines for Maritime Interaction, the objectives of the Guidelines include the following:
    4.1. To advance ASEAN's maritime security efforts with the end view of realizing the goals of ASEAN defence ministers.
    4.2. To establish comprehensive and feasible maritime conflict management measures on the basis of confidence-building, preventive diplomacy and peaceful management of tensions that could arise at sea.
    4.3. To contribute in addressing common maritime security challenges faced by ASEAN member states.

4.4. To contribute to the implementation of international law and regional conventions including, among others, the United Nations Convention on the Law of the Sea (UNCLOS), International Regulations for Preventing Collisions at Sea (COLREG) and CUES.

4.5. To serve as a set of guidelines for the ASEAN defence sectoral body in engaging other relevant ASEAN sectoral bodies involved in maritime security.

Scope and application

5. The end users of the Guidelines will primarily be ASEAN defence establishments, particularly naval ships and naval aircraft.

6. The Guidelines shall only apply when the subject naval ships or naval aircraft are from ASEAN member states.

7. The Guidelines shall be applicable when the subject naval vessels are in the high seas.

8. The implementation of the Guidelines shall be voluntary and non-legally binding.

   It also does not create any international obligation or commitment under international law.

9. The possibility of extending guidelines with the Plus countries shall only be explored and decided by the ADMM.

10. In the event that there is a decision to extend the Guidelines with Plus countries, this shall be based on consensus of the ADMM upon the endorsement of the ASEAN Defence Senior Officials' Meeting (ADSOM) through the ADSOM Working Group (WG).

General principles

11. The Guidelines shall be based on ASEAN's fundamental principles as set out in Article 2 of the ASEAN Charter.

12. The Guidelines shall uphold all existing maritime arrangements of ASEAN member states. It shall not supersede any international agreement or treaty.

13. The Guidelines shall reaffirm the ASEAN member states' commitment to resolve disputes through peaceful means without resorting to the threat or use of force in accordance with universally recognised principles of international law, including UNCLOS.

14. The Guidelines are without prejudice to: (i) existing rights and obligations of both user and coastal states under international law, including UNCLOS; (ii) existing rights and obligations under bilateral and multilateral arrangements between states, as well as between states and organisations and (iii) ASEAN member states' positions vis-à-vis existing maritime and airspace disputes.

Definition of terms

15. A naval vessel refers to warships as defined by UNCLOS, naval auxiliaries as defined by CUES, and submarines.
16. A naval aircraft refers to fixed-wing and rotary-wing aircraft, and unmanned aerial systems or vehicles that are used by the armed forces of a state in maritime operations.
17. The definitions provided by UNCLOS on different maritime zones, including internal waters, archipelagic waters, exclusive economic zone, continental shelf and high seas, shall be followed in this Guidelines.

Interaction with naval ships and aircraft

18. Foreign naval ships enjoy certain immunities in accordance with international law.
19. Preservation of life and property should be of utmost consideration.
20. Naval ships and naval aircraft presenting a challenge should be warned and given the opportunity to withdraw or otherwise cease its actions.
21. Upon issuance of a query or warning from a naval ship or naval aircraft of another ASEAN member state, the naval ship or naval aircraft in question should identify itself.
22. When calls are initiated, naval ships and naval aircraft are encouraged to promptly respond to avoid miscalculations or misunderstanding.
23. When such miscalculations or misunderstanding occur, naval ships and naval aircraft should increase efforts to communicate.
24. ASEAN member states should follow communication procedures anchored on CUES.
25. In the absence of a perceived insecurity, naval ships of ASEAN member states within each other's line of sight are encouraged to exchange information through the Automatic Identification System (AIS).
26. During unplanned encounters at sea, naval ships are encouraged to conduct passing exercises and communications exercises.
27. Naval ships and naval aircraft may refer to relevant provisions in CUES and COLREG to avoid untoward incidents, particularly on safe speed, safe distance, assurance measures and signals.
28. In the event of an untoward incident, subject naval ships should refrain from taking any action that will further escalate the situation. Efforts should focus on rescue of personnel as required by international law and in line with the capacity of the naval ship or naval aircraft. One ship or aircraft may not, however, board or salvage the ship or aircraft of the other side without prior explicit consent.
29. During peacetime, naval ships are encouraged to turn on the AIS in high-traffic areas to avoid untoward incidents.

Rendering assistance

30. Should an extreme emergency arise that indicate the need for assistance to preserve a life, nearby naval ships and naval aircraft should endeavour to extend assistance upon the request of the distressed naval ship.
31. When requesting assistance, the distressed naval ship should provide all necessary information, including the patient's condition, weather condition, as well as the ship's accurate position, time, speed and course. In case of an aircraft transfer from a naval ship, the aircraft should be informed of the hoist location.

Interaction with civilian maritime agencies

32. The ASEAN defence sectoral body is encouraged to engage other relevant ASEAN sectoral bodies involved in maritime security to enhance interoperability and promote cross-pillar cooperation.
33. The convening of an expanded ad hoc working group composed of policy and technical officials from ASEAN member states' defence establishments and other maritime security agencies may be initiated by the defence establishment of any ASEAN member state on a voluntary basis for the purposes of sharing knowledge, experiences and best practices, and exploring opportunities for cooperation to avoid untoward incidents at sea, including the possible expansion of the Guidelines to relevant civilian agencies.
34. Extending the Guidelines to other ASEAN sectoral bodies shall only be considered once the Guidelines for Maritime Interaction has been finalised and tested within the ASEAN defence sector.

Synergy with other related efforts

35. During maritime-related emergencies that require timely communication and decision-making between ASEAN defence ministers, the ASEAN Direct Communications Infrastructure (ADI) should remain as the primary mechanism for 'providing a permanent, rapid, reliable, and confidential means by which any two ASEAN Defence Ministers may communicate with each other to arrive at mutual decisions in handling crisis or emergency situations, in particular related to maritime security' as reflected in the Concept Paper Establishing a Direct Communications Link in the ADMM Process.
36. The Guidelines should be a complementary initiative for naval ships and aircraft alongside the Guidelines for Air Military Encounters (GAME).
37. Relevant outcomes of related meetings, namely workshops, seminars, exercises and other activities, under the ADMM-Plus Experts' Working Group (EWG) on Maritime Security as well as those from the ASEAN Regional Forum (ARF), East Asia Summit (EAS), ASEAN Maritime Forum (AMF)

and Expanded ASEAN Maritime Forum (EAMF), among others, may be taken into consideration to provide inputs for the development and implementation of the Guidelines.

38. Other initiatives that should also be considered are those by the Western Pacific Naval Symposium (WPNS).

Implementation and amendments

39. An ad hoc working group composed of policy and technical officials from the defence establishments of ASEAN member states should be established to monitor the development and implementation of the Guidelines, as well as developments in other related initiatives. Relatedly, any ASEAN member state that hosted a similar initiative should bring such initiatives to the attention of the ad hoc working group.

40. The ASEAN Navy Chiefs' Meeting (ANCM) shall be the lead body for formulating and developing the operational and technical parameters of the Guidelines. Feedback and status of the implementation of the Guidelines shall be reported by the ANCM to the ad hoc working group for onward submission to the ADSOM through the ADSOM WG.

41. The outcome of the meetings and workshops as well as the status of the development and implementation of the Guidelines shall be duly reported to and assessed by the ADMM through the ADSOM and ADSOM WG.

42. The Guidelines shall be considered as a living document that can be amended based on the consensus of the ADMM through the ADSOM and ADSOM WG.

43. Proposed amendments shall be presented by the ad hoc working group to the ADSOM WG for discussion and deliberation. Once a consensus is reached, the amended Guidelines shall be submitted to the ADSOM for endorsement to the ADMM for adoption.

References

Ackerman, E. and A. Silver (2016). 'This Robot Can Fly a Plane From Takeoff to Landing'. 15 November 2016, *IEEE Spectrum*.

Affairs, A. I. o. I. (2017). 'Julie Bishop Speaks at AIIA 2017 National Conference'.

Affairs, B. o. P.-M. (2017). 'Joint Declaration for the Export and Subsequent Use of Armed or Strike-Enabled Unmanned Aerial Vehicles (UAVs)'. 16 October 2017, *Fact Sheet*, https://www.state.gov/t/pm/rls/fs/2017/274817.html.

Affairs, M. o. H. (2017). 'Singapore Terrorism Threat Assessment Report 2017'.

Affairs, M. o. H. (2019). 'Singapore Terrorism Threat Assessment Report 2019'.

AFP (2018). 'Indonesia Inks US$1.1b Deal with Russia to Buy 11 Sukhoi Jets'. 17 February 2018, *New Straits Times*.

Agastia, I. G. B. D. (2017). 'Small Navy, Big Responsibilities: The Struggles of Building Indonesia's Naval Power'. *AEGIS* 1, 2.

Allen, G. C. (2019). *Understanding China's AI Strategy: Clues to Chinese Strategic Thinking on Artificial Intelligence and National Security*. Centre for a New American Security.

Allison, G. T. (2017). *Destined for War: Can America and China Escape Thucydide's Trap?*. Scribe Publications.

Almanar, A. (2017). 'Govt to Issue Integrated Weapons Regulation After TNI vs. Police Quarrel'. 14 October 2017, Jakarta Globe.

Altmann, J. and F. Sauer (2017). 'Autonomous Weapon Systems and Strategic Stability'. *Survival* 59(5), 117–142.

Anderson, K. (2016). 'Why the Hurry to Regulate Autonomous Weapon Systems—But Not Cyber-Weapons'. *Temple International and Comparative Law Journal* 30(1), 17–42.

Anderson, K., D. Reisner and M. C. Waxman (2014). 'Adapting the Law of Armed Conflict to Autonomous Weapon Systems'. *International Law Studies* 90, 386–411.

Andersson, J. J. (2015). 'Submarine Capabilities and Conventional Deterrence in Southeast Asia'. *Contemporary Security Policy* 36(3), 473–497.

Antey, A. (2018). 'US to Exempt India, Indonesia and Vietnam From CAATSA Sanctions'. 24 July 2018, *DefenseWorld.net*.

Anthony, I., L. Grip and C. Holland (2014). 'The Governance of Autonomous Weapons'. *SIPRI Yearbook*, Oxford: Oxford University Press.

Appier (2018). 'Artificial Intelligence Is Critical to Accelerate Digital Transformation in Asia Pacific'. *Forrester Opportunity Snapshot*, Forrester.

Arif, M. and Y. Kurniawan (2018). 'Strategic Culture and Indonesian Maritime Security'. *Asia and the Pacific Policy Studies* 5(1), 77–89.

Arkin, R. C. (2008). 'Governing Lethal Behavior: Embedding Ethics in a Hybrid Deliberative/Reactive Robot Architecture Part I: Motivation and Philosophy'. 2008 3rd ACM/IEEE International Conference on Human-Robot Interaction (HRI), IEEE.

Arkin, R. C. (2010). 'The Case for Ethical Autonomy in Unmanned Systems'. *Journal of Military Ethics* 9(4), 332–341.

Arkin, R. C. (2013). 'Lethal Autonomous Systems and the Plight of the Non-Combatant'. *AISB Quarterly* 137, 1–9.

Asaro, P. (2008). 'How Just Could a Robot War Be'. *Proceedings of the 2008 Conference on Current Issues in Computing and Philosophy*, 50–64.

Asaro, P. (2019). 'What Is an Artificial Intelligence Arms Race Anyway'. *ISJLP* 15, 45.

Asaro, P. M. (2008). 'How Just Could a Robot War Be'. In: *Current Issues in Computing and Philosophy* PE Brey, A Briggle, and K Waelbers (eds.). IOS Press, pp. 50–64.

Asaro, P. M. (2016). 'The Liability Problem for Autonomous Artificial Agents'. *Ethical and Moral Considerations in Non-Human Agents, 2016 AAAI Spring Symposium Series.*

Asia, B. (2018, 9 January). 'Indonesia Country Profile'. Retrieved 27 February 2018, from http://www.bbc.com/news/world-asia-pacific-14921238.

Asia, J. s. S. S. A.-S. (2018). 'Defence Production and R&D'. Jane's By IHS Markit.

Asia, J. s. S. S. A.-S. (2018). 'Singapore—Armed Forces'. Jane's by IHS Markit; Asia, J. s. S. S. A.-S. (2018). 'Defence Production and R&D'. Jane's By IHS Markit.

Atsma, A. J. (2000). 'Automotons (Automotones)'. *Theoi Project.* http://www.theoi.com/Ther/Automotones.html.

Ba, A. D. (2017). 'ASEAN and the Changing Regional Order: The ARF, ADMM, and ADMM-Plus'. In: *Building ASEAN Community: Political–Security and Socio-Cultural Reflections* A. Baviera and L. Maramis (eds.). Economic Research Institute for ASEAN and East Asia, Jakarta, Indonesia, 146.

Baharudin, H. (2019). 'Digital Defence to Be Sixth Pillar of Total Defence'. 15 February 2019, *The Straits Times.*

Baker, J. (2019). 'President Trump's Executive Order on Artificial Intelligence'. 28 February 2019, Lawfare.

Bank, T. W. (2017). 'Gross Domestic Product 2017'. https://databank.worldbank.org/data/download/GDP.pdf.

Bank, T. W. (2017). 'Military Expenditure (% of GDP)'. https://data.worldbank.org/indicator/MS.MIL.XPND.GD.ZS.

Bank, T. W. (2017). 'Overview'. 19 September 2017, The World Bank in Indonesia. Retrieved 04 March 2018, from http://www.worldbank.org/en/country/indonesia/overview.

Bank, T. W. (2017, 19 September). 'Overview'. The World Bank in Indonesia. Retrieved 04 March 2018, from http://www.worldbank.org/en/country/indonesia/overview.

Bank, T. W. (2019). Databank: Indonesia.

Barkawi, T. and M. Laffey (2006). 'The Postcolonial Moment in Security Studies'. *Review of International Studies* 32(2), 329–352.

Bars, P. (2015). *Asia-Pacific Defence Outlook 2015: Tension, Collaboration, Convergence.* Deloitte.

Bevins, V. (2017). 'In Indonesia, the 'Fake News' that Fueled a Cold War Massacre is Still Potent Five Decades Later'. 30 September 2017, The Washington Post.

Bin Osman, M. M. (2014). *Speech by Minister of State for Defence, Dr Mohamad Maliki Bin Osman, at the Young Defence Scientists Programme Congress 2014.* Ministry of Defence.

Binnie, J. (2018). 'Russians Reveal Details of UAV Swarm Attacks on Syrian Bases'. 12 January 2018, *IHS Jane's Defence Weekly*.

Bisley, N. (2017). 'The East Asia Summit and ASEAN: Potential and Problems'. *Contemporary Southeast Asia: a Journal of International and Strategic Affairs* 39(2), 265–272.

Bitzinger, R. (2015). 'IMDEX ASIA: Southeast Asian Naval Expansion and Defence Spending'. *RSIS Commentary*.

Bitzinger, R. (2018). 'Military-Technological Innovation in Small States: The Cases of Israel and Singapore'. *SITC Research Briefs* 10(4), 1–4.

Bitzinger, R. A. (2010). 'A New Arms Race? Explaining Recent Southeast Asian Military Acquisitions'. *Contemporary Southeast Asia* 32(1), 50–69.

Bitzinger, R. A. (2013). 'Revisiting Armaments Production in Southeast Asia: New Dreams, Same Challenges'. *Contemporary Southeast Asia* 35(3), 369–394.

Bitzinger, R. A. (2017). 'Asian Arms Industries and Impact on Military Capabilities'. *Defence Studies* 17(3), 295–311.

Blaxland, J. (2015). 'Australia's 1999 Mission to East Timor part 1: The Decision to Intervene'. *The Strategist*. https://www.aspistrategist.org.au/australias-1999-mission-to-east-timor-part-1-the-decision-to-intervene/.

Boot, M. (2006). *War Made New: Technology, Warfare, and the Course of History, 1500 to Today*. Gotham Books.

Boulanin, V. and M. Verbruggen (2017). *Mapping the Development of Autonomy in Weapon Systems*. Stockholm International Peace Research Institute.

Brown, M. and P. Singh (2018). *China's Technology Transfer Strategy: How Chinese Investments in Emerging Technology Enable A Strategic Competitor to Access the Crown Jewels of U.S. Innovation*. D. I. U. Experimental.

Bunker, R. J. (2015). *Terrorist and Insurgent Unmanned Aerial Vehicles: Use, Potentials, and Military Implications*. Carlisle, PA: US Army Strategic Studies Institute.

Bureau (2019). 'Russia's New Stealth Drone May Operate Together With Su-57 Jet'. 26 August 2019, *Defenseworld.net*.

Burke, A. E. (2017). *Torpedoes and Their Impact on Naval Warfare*. Fort Belvoir: Defense Technical Information Center.

Butcher, J. G. (2013). 'The International Court of Justice and the Territorial Dispute between Indonesia and Malaysia in the Sulawesi Sea'. *Contemporary Southeast Asia: a Journal of International and Strategic Affairs* 35(2), 235–257.

Canning, J. S. (2009). 'You've Just Been Disarmed. Have a Nice Day!'. *IEEE Technology and Society Magazine* 28(1), 13–15.

Carpenter, C. (2013). 'US Public Opinion on Autonomous Weapons'. *Duck of Minerva Blog*. http://duckofminerva.dreamhosters.com/wp-content/uploads/2013/06/UMass-Survey_Public-Opinion-on-Autonomous-Weapons.pdf.

Carroll, D., H. Everett, G. Gilbreath and K. Mullens (2002). *Extending Mobile Security Robots to Force Protection Missions*. DTIC Document.

Carter, W. (2018). 'Statement Before the House Armed Services Committee Subcommittee on Emerging Threats and Capabilities—"Chinese Advances in Emerging Technologies and Their Implications for U.S. National Security"'. *House Armed Services Committee Subcommittee on Emerging Threats and Capabilities*. Rayburn House Office Building.

Carter, W. A., E. Kinnucan, J. Elliot, W. Crumpler and K. Lloyd (2018). 'A National Machine Intelligence Strategy for the United States'. *CSIS Technology Policy Program*, Center for Strategic & International Studies.

Cassingham, G. J. (2016). *Remotely Effective: Unmanned Aerial Vehicles, The Information Revolution In Military Affairs, And The Rise Of The Drone In Southeast Asia '. Master of Arts in Security Studies (Far East, Southeast Asia, the Pacific)*. Naval Postgraduate School.

Caverley, J. D. (2017). 'Slowing the Proliferation of Major Conventional Weapons: The Virtues of an Uncompetitive Market'. *Ethics and International Affairs* 31(4), 401–418.

CCW, C. D. t. (2018). Position Paper.

Center, A. C. I. (2017). *Robotic and Unmanned Systems Strategy*. U.S. Army Training and Doctrine Command.

Center, P. R. (2014). 'Global Opposition to U.S. Surveillance and Drones, but Limited Harm to America's Image'. 14 July 2014, *Global Attitudes and Trends*. http://www.pewglobal.org/2014/07/14/global-opposition-to-u-s-surveillance-and-drones-but-limited-harm-to-americas-image/.

Center, P. R. (2014, 14 July). 'Global Opposition to U.S. Surveillance and Drones, but Limited Harm to America's Image'. *Global Attitudes and Trends*. http://www.pewglobal.org/2014/07/14/global-opposition-to-u-s-surveillance-and-drones-but-limited-harm-to-americas-image/.

Centre, P. G. M. (2018). *The Future of ASEAN - Time to Act, PwC*.

Chairil, T. (2018). 'A Self-Reliant Defence Industry: A Mission Impossible for Indonesia?'. 3 July 2018, *The Conversation*.

Chairil, T. (2018). 'The Politics behind Alpalhankam: Military and Politico-Security Factors in Indonesia's Arms Procurements, 2005–2015'. In: *Competition and Cooperation in Social and Political Sciences* I. R. Adi and R. Achwan (eds.). Routledge, 281–290.

Chamayou, G. (2011). 'The Manhunt Doctrine'. *Radical Philosophy* 169, 2–6.

Chamayou, G. (2015). *Drone Theory*, Penguin Books Limited.

Chan, J. (2016). 'Singapore and the South China Sea: Being an Effective Coordinator and Honest Broker'. *Asia Policy* 21(1), 41–46.

Chandran, N. (2017). 'Beijing is Using Underwater Drones in the South China Sea to Show Off Its Might'. 12 August 2017, CNBC.

Chase, M. S., K. A. Gunness, L. J. Morris, S. K. Berkowitz and B. S. Purser III (2015). *Emerging Trends in China's Development of Unmanned Systems*. Santa Monica: RAND National Defense Research Institute.

Cheater, J. C. (2007). Accelerating the Kill Chain via Future Unmanned Aircraft. *CFS a. T A w. College*.

Chitturu, S., D.-Y. Lin, K. Sneader, O. Tonby and J. Woetzel (2017). 'Artificial Intelligence and Southeast Asia's Future'. *Discussion Paper*. McKinsey & Company.

Chivers, C. J. (2010). *The Gun: The Story of the AK-47*. Penguin Books Limited.

Cho, H.-B. (2018). 'Tying the Adversary's Hands: Provocation, Crisis Escalation, and Inadvertent War'. PhD Dissertation, University of Pennsylvania.

Choy, D., K. Ju-Hon, L. C. Han, L. S. Hiang, J. Leong, R. Ng and F. Teo (2003). 'Creating the Capacity to Change: Defence Entrepreneurship for the 21st Century'. Monograph.

Christensen, C. M. (2015). *The Innovator's Dilemma: When New Technologies Cause Great Firms to Fail*. Harvard Business Review Press.

Coeckelbergh, M. (2013). 'Drones, Information Technology, and Distance: Mapping the Moral Epistemology of Remote Fighting'. *Ethics and Information Technology* 15(2), 87–98.

Collins, J., and A. Futter. (2015). 'Introduction: Reflecting on the Global Impact of the RMA'. In: *Reassessing the Revolution in Military Affairs*. Springer, London, 1–15.

Conn, A. (2016). 'The Problem of Defining Autonomous Weapons'. Future of Life Institute. https://futureoflife.org/2016/11/30/problem-defining-autonomous-weapons/.

Cooper, H. (2016). 'U.S. Demands Return of Drone Seized by Chinese Warship'. *The New York Times*.

Cornillie, C. (2018). 'Can Pentagon Bridge Artificial Intelligence's 'Valley of Death'?'. *Bloomberg Government*. https://about.bgov.com/news/can-pentagon-bridge-artificial-intelligences-valley-of-death/?utm_source=newsletter&utm_medium=email&utm_c ampaign=newsletter_axiosfutureofwork&stream=future.

Council, U. N. S. (2016). '7758th Meeting: Non-Proliferation of Weapons of Mass Destruction'.

Crime, U. N. O. o. D. a. (2016). *Protecting Peace and Prosperity in Southeast Asia: Synchronizing Economic and Security Agendas*. United Nations Office on Drugs and Crime.

Cronin, P. M. and S. Lee (2017). 'Expanding South Korea's Security Role in the Asia-Pacific Region'. *Discussion Paper*. Council on Foreign Relations. Retrieved from https://www.cfr.org/sites/default/files/pdf/2017/03/Discussion_Paper_Cronin_Lee_China_ROK_Security_OR.pdf

Cronk, T. M. (2018). 'Artificial Intelligence Experts Address Getting Capabilities to Warfighters'. 12 February 2019, *United States Department of Defense*.

Crootof, R. (2014). 'The Killer Robots Are Here: Legal and Policy Implications'. *Cardozo Law Review* 36, 3–51.

Crootof, R. (2016). 'A Meaningful Floor for 'Meaningful Human Control''. *Temple International and Comparative Law Journal* 30, 1.

Crootof, R. (2016). 'War Torts: Accountability for Autonomous Weapons'. *University of Pennsylvania Law Review* 164, 6.

D. Stork, MIT Press, Guetlein, M. A. (2005). *Lethal Autonomous Weapons--Ethical and Doctrinal Implications*. DTIC Document.

Davies, P. (2015). 'The ADF and Armed Drones'. *The Strategist*. http://www.aspistrategist.org.au/the-adf-and-armed-drones/.

Defence, D. o (2012). Directive 3000.09.

Defence, M. o (2017). *Defence Technology Is Key Enabler for Next Gen SAF*. Singapore.

Defence, M. o (2017, 27 December). 'Defence Spending'. *MINDEF Policies*. https://www.mindef.gov.sg/web/portal/mindef/defence-matters/mindef-policies/mindef-policies-detail/defence-spending.

Defence, M. o (2019). *Reply to TODAY's Query on Status of MINDEF/SAF's F-35 Acquisition Following Japanese F-35 Crash*. Singapore.

Defence, S. M. o (2018). *Crown Prince of Brunei Visits the RSAF's Unmanned Aerial Vehicle Command*. Singapore Ministry of Defence.

Defence, U. K. M. o (2011). *Joint Doctrine Note 2/11: The UK Approach to Unmanned Aircraft Systems*. U. K. M. o. Defence.

Defense, D. o (2019). *Summary of the 2018 Department of Defense Artificial Intelligence Strategy: Harnessing AI to Advance Our Security and Prosperity*. Department of Defense.

Defense, O. o. t. S. o (2015). *Military and Security Developments Involving the People's Republic of China 2015*. Department of Defence.

Dennet, D. (1996). 'When HAL Kills, Who's to Blame?'. *HAL's Legacy: 2001'S Computer as Dream and Reality*. D. Stork, MIT Press.

Desk, N. (2018). 'Indonesia, Turkey Team up to Develop Military Drones'. *The Jakarta Post*, West Java.

Desker, B. and R. A. Bitzinger (2016). 'A Perspective on Singapore'. *Proliferated Drones*, Center for a New American Security.

Development, C. a. D. C. (2011). 'Joint Doctrine Note 2/11: The UK Approach To Unmanned Aircraft Systems'. *Joint Doctrine Note*, U.K. Ministry of Defence.

Development, C. a. D. C. (2017). 'Joint Doctrine Publication 0-30.2: Unmanned Aircraft Systems'. *Joint Doctrine Note*, U.K. Ministry of Defence.

Development, C. a. D. C. (2018). 'Joint Concept Note 1/18 Human Machine Teaming'. *Joint Concept Note*, U.K. Ministry of Defence.

Dexian, C. C. (2013). 'Hedging for Maximum Flexibility: Singapore's Pragmatic Approach to Security Relations with the US and China'. *Pointer: Journal of the Singapore Armed Forces* 39, 1–12.

DiCicco, J. M. and J. S. Levy (1999). 'Power Shifts and Problem Shifts: The Evolution of the Power Transition Research Program'. *Journal of Conflict Resolution* 43(6), 675–704.

Dinstein, Y. (2016). *The Conduct of Hostilities Under the Law of International Armed Conflict.* Cambridge: Cambridge University Press.

Diprose, R., D. McRae and V. R. Hadiz (2019). 'Two Decades of Reformasi in Indonesia: Its Illiberal Turn'. *Journal of Contemporary Asia*, 1–22.

Division, P. S. (2015). 'Securing Singapore: From Vulnerability to Self-Reliance'. https://www.psd.gov.sg/heartofpublicservice/our-institutions/securing-singapore-from-vulnerability-to-self-reliance/.

Docherty, B. (2012). *Losing Humanity: The Case Against Killer Robots.* Human Rights Watch.

Donald, D. (2014). 'Wulung UAV Gets Stronger and Lighter'. 06 November 2014, Jane's 360.

Dowdy, J., D. Chinn, M. Mancini and J. Ng (2014). *Southeast Asia: The Next Growth Opportunity in Defense.* McKinsey Innovation Campus Aerospace and Defense Practice.

Doyle, J. S. (2016). 'The Yom Kippur War and the Shaping of the United States Air Force'. Masters, Air University.

Drone, C. f. t. S. o. t. (2016). 'The Drone Database'. Retrieved 20 July 2017, from http://drones.cnas.org/drones/.

Economic Policy Group. (2018). Survey of professional forecasters, Monetary Authority of Singapore, https://www.mas.gov.sg/-/media/MAS/EPG/SPF/2018/MAS-SPF-Documentation-2018.pdf

Edelstein, S. (2018). 'Hyundai Tests Autonomous Semitruck Technology on South Korean Highway'. *The Drive.*

Education, C. E. L. C. f. D. D. a. (2017). 'Annex 3-60 Targeting'. Command. United States Air Force Air Education and Training Command, Maxwell Air Force Base: Montgomery, Alabama.

Ellman, J., L. Samp and G. Coll (2017). *Assessing the Third Offset Strategy.* Center for Strategic & International Studies.

Engineering, C. f. T. a. I. (2003). 'Remote Piloted Aerial Vehicles : An Anthology'. *Aviation and Aeromodelling: Interdependent Evolutions and Histories.* Retrieved 20 December 2017, from http://www.ctie.monash.edu.au/hargrave/rpav_home.html.

Everett, H. R. and M. Toscano (2015). *Unmanned Systems of World Wars I and II.* MIT Press.

Ewers, E. C., L. Fish, M. C. Horowitz and P. Scharre (2017). 'Drone Proliferation: Choices for the Trump Administration'. Papers for the President, Centre for a New American Security.

Ewers, E. C., L. Fish, M. C. Horowitz, A. Sander and P. Scharre (2017). 'Drone Proliferation: Policy Choices for the Trump Administration'. Papers for the President, Center for a New American Security.

Excellence, U. S. A. U. C. o (2010). *Eyes of the Army: U.S. Army Unmanned Aircraft Systems Roadmap 2010–2035'. Fort Rucker*. U.S. Army.

Factbook, C. W. (2019, 7 May). 'Singapore'. https://www.cia.gov/library/publications/resources/the-world-factbook/geos/sn.html.

Farley, R. M. (2018). 'South Korea's Second Dokdo-Class Assault Carrier and the Future of the ROKN'. *The Diplomat*. https://thediplomat.com/2018/05/south-koreas-second-dokdo-class-assault-carrier-and-the-future-of-the-rokn/.

Farley, R. M. and Y. Gortzak (2009). 'Fighting Piracy: Experiences in Southeast Asia and off the Horn Of Africa'. *Journal of Strategic Security* 2, 1.

Fels, E. (2017). *Shifting Power in Asia-Pacific? The Rise of China, Sino-US Competition and Regional Middle Power Allegiance*. Springer International Publishing.

Figueroa, A. (2018). 'License to Kill: An Analysis of the Legality of Fully Autonomous Drones in the Context of International Use of Force Law'. *Pace International Law Review* 31, 145.

Finley, K. (2016). 'AI Fighter Pilot Beats A Human, but No Need to Panic (Really)'. 29 June 2016, *Wired*.

Finn, A. and S. Scheding (2012). *Developments and Challenges for Autonomous Unmanned Vehicles*. Springer.

Finnemore, M. and J. Goldstein (2013). *Back to Basics: State Power in a Contemporary World*. Oxford: Oxford University Press.

Fleurant, A., A. Kuimova, N. Tian, P. D. Wezeman and S. T. Wezeman (2018). 'The SIPRI Top 100 ArmsProducing and Military Services Companies, 2017'. *Sipri Fact Sheet*. Stockholm International Peace Research Institute.

Fleurant, A., P. D. Wezeman, S. T. Wezeman and N. Tian (2017). 'Trends in International Arms Transfers, 2016'. *SIPRI Fact Sheet*. Retrieved from https://www.sipri.org/sites/default/files/Trends-in-international-arms-transfers-2016.pdf.

Fook, L. L. (2018). 'Singapore–China Relations: Building Substantive Ties amidst Challenges'. *Southeast Asian Affairs* 2018, 321–339.

Forbes (2018). 'Singapore'. *Best Countries for Business*. https://www.forbes.com/places/singapore/.

Freedman, L. (2017). *The Future of War: A History*. Penguin Books Limited.

Fund, I. M. (2019). *World Economic Outlook Database*.

Gady, F.-S. (2015). 'Vietnam Reveals New Drone for Patrolling the South China Sea'. *The Diplomat*.

Galdi, T. (1995). 'Revolution in Military Affairs'. *CRS Report for Congress*.

Galliott, J. (2015). *Military Robots: Mapping the Moral Landscape*. Ashgate Publishing, Ltd.

Galliott, J. C. (2012). 'Uninhabited Aerial Vehicles and the Asymmetry Objection: A Response to Strawser'. *Journal of Military Ethics* 11(1), 58–66.

Garfinkel, B. and A. Dafoe (2019). 'How Does the Offense-Defense Balance Scale?' *Journal of Strategic Studies* 42(6), 736–763.

Gaub, D. L. (2011). *Children of Aphrodite: The Proliferation and Threat of Unmanned Aerial Vehicles in the Twenty-First Century*. DTIC Document.

Geneva, T. U. N. O. a. (2019). *Report of the 2019 Session of the Group of Governmental Experts on Emerging Technologies in the Area of Lethal Autonomous Weapons Systems (Advance Version)*. The Convention on Certain Conventional Weapons.

Geneva Academy (2014). *Academy Briefing 8: Autonomous Weapon Systems under International Law*. Geneva: Geneva Academy of International Humanitarian Law and Human Rights.

Gettinger, D. (2016). *Drone Spending in the Fiscal Year 2017 Defense Budget*. New York: Centre for the Study of the Drone at Bard College.

Gibbs (2015). 'Musk, Wozniak and Hawking Urge Ban on Warfare AI and Autonomous Weapons'. 27 July 2015, *The Guardian*.

Gilli, A. and M. Gilli (2014). 'The Spread of Military Innovations: Adoption Capacity Theory, Tactical Incentives, and the Case of Suicide Terrorism'. *Security Studies* 23(3), 513–547.

Gilpin, R. (1988). 'The Theory of Hegemonic War'. *Journal of Interdisciplinary History* 18(4), 591–613.

Goh, G. (2003). 'The 'ASEAN Way': Non-Intervention and ASEAN's Role in Conflict Management'. *Stanford Journal of East Asian Affairs* 3, 1.

Goldman, E. and T. Mahnken (2004). *The Information Revolution in Military Affairs in Asia*. United States: Palgrave Macmillan.

Goldman, E. O. and R. B. Andres (1999). 'Systemic Effections of Military Innovation and Diffusion'. *Security Studies* 8(4), 79–125.

Goldman, E. O. and L. C. Eliason (2003). *The Diffusion of Military Technology and Ideas*. Stanford University Press.

Gramer, R. (2017). 'Afghan Insurgents Use Drones in Fight Against U.S.'. *Foreign Policy*.

Grant, H. (2015). 'UN Delay Could Open Door to Robot Wars, Say Experts'. 6 October 2015.

Gregg, A. (2019). *Autonomous Police Vehicles: The Impact on Law Enforcement*. Monterey: DTIC. N. P. School.

Gregory, D. (2011). 'From a View to a Kill Drones and Late Modern War'. *Theory, Culture and Society* 28(7–8), 188–215.

Grevatt, J. (2017). 'Korean-Indonesian Fighter Project Hits Licensing Delays'. *IHS Jane's Defence Industry*.

Grevatt, J. (2018). 'Indonesia Enacts Law to Boost Collaboration with South Korea. Jane's'. *Jane's 360*, Jane's Defence Industry.

Grevatt, J. (2018). 'Indonesia Registers USD284 Million in Defence Exports'. 23 November 2018, *Jane's Defence Industry*.

Grevatt, J. (2018). 'Smart Moves: Fourth Industrial Revolution Technologies in Asia'. 21 December 2018, *Jane's Defence Weekly*, IHS Markit.

Grevatt, J. (2018). *South Korean Military Exports Climb 25%*. IHS Jane's Defence Industry.

Grevatt, J. and R. Rahmat (2018). *Indonesia's Defence Market Poised to Expand*. IHS Markit.

Grissom, A. (2006). 'The Future of Military Innovation Studies'. *Journal of Strategic Studies* 29(5), 905–934.

Gross, J. A. (2017). 'IDF Company Commanders to Receive Collapsible Drones by Year's End'. 1 June 2017, *Times of Israel*.

Grossman, D. and L. W. Christensen (2007). *On Combat: The Psychology and Physiology of Deadly Conflict in War and in Peace*. Belleville, IL: PPCT Research Publications.

Group, E. P. (2018). *Macroeconomic Review*. Monetary Authority of Singapore.

Gunaratna, R. (2017). 'The Changing Threat Landscape: Countering Terrorism in Singapore'. In: *The Palgrave Handbook of Global Counterterrorism Policy*

S. N. Romaniuk, F. Grice, D. Irrera and S. Webb (eds.). United Kingdom: Palgrave Macmillan, 749–769.

Haacke, J. (2009). 'The ASEAN Regional Forum: from Dialogue to Practical Security Cooperation?' *Cambridge Review of International Affairs* 22(3), 427–449.

Håland, W. C. (2018). 'Weaponized Multi-Utility Unmanned Ground Vehicles'. *Small Arms Defense Journal*. http://www.sadefensejournal.com/wp/weaponized-multi-uti lity-unmanned-ground-vehicles/.

Hallett, A. and V. Weedn (2016). 'Unmanned Systems Technology Use by Law Enforcement'. In: *Forensic Science: A Multidisciplinary Approach* E. Katz and J. Halámek (eds.). John Wiley & Sons Incorporated.

Hamilton, E. J. and B. C. Rathbun (2013). 'Scarce Differences: Toward a Material and Systemic Foundation for Offensive and Defensive Realism'. *Security Studies* 22(3), 436–465.

Hammond, D. N. (2014). 'Autonomous Weapons and the Problem of State Accountability'. *Chi. J. Int'l L.* 15, 652.

Hanson, F., T. Uren, F. Ryan, M. Chi, J. Viola and E. Chapman (2017). *Cyber Maturity in the Asia Pacific Region 2017.* Australian Strategic Policy Institute.

Hardy, J. (2013). 'Rheinmetall Confirms Indonesian Leopard 2 Contract'. 13 November 2013, *Jane's Defence Weekly.*

Haripin, M. (2016). 'Rearming the Indonesian State: The Role of Defence Industry Policy Committee'.立命館国際地域研究 44(12), 39–58.

Harris, B. (2016). 'Global Instability Drives S Korean War Industry'. *Financial Times.*

Harwell, D. (2018). 'Defense Department Pledges Billions Toward Artificial Intelligence Research'. 7 September 2018, *The Washington Post.*

Haryanto, J. T. (2017). 'Potential Impact of the Fulfilment of Minimum Essential Force (MEF) to the Regional Welfare'. *Jurnal Ekonomi Kuantitatif Terapan* 10, 2.

Heather Roff Quoted in Conn, A. (2016). 'The Problem of Defining Autonomous Weapons'. *Future of Life Institute.* https://futureoflife.org/2016/11/30/problem-defini ng-autonomous-weapons/.

Heiduk, F. (2017). 'An Arms Race in Southeast Asia? Changing Arms Dynamics, Regional Security and the Role of European Arms Exports'. *SWP Research Paper*, Stiftung Wissenschaft und Politik.

Hellyer, M. (2019). *The Cost of Defence: ASPI Defence Budget Brief 2019–20.* Canberra: Australia: Australian Strategic Policy Institute.

Hermansyah, A. (2016). 'Shooting for the Moon: Eyeing the World's Best Weapons Store Industry'. 08 August 2018, *The Jakarta Post.*

Heyns, C. (2013). *Report of the Special Rapporteur on Extrajudicial, Summary or Arbitrary Executions (A/HRC/23/47).* United Nations General Assembly.

Heyns, C. (2016). 'Human Rights and the Use of Autonomous Weapons Systems (AWS) during Domestic Law Enforcement'. *Human Rights Quarterly* 38(2), 350–378.

Heyns, C. (2017). 'Autonomous Weapons in Armed Conflict and the Right to a Dignified Life: An African Perspective'. *South African Journal on Human Rights* 33(1), 46–71.

Hockstein, N. G., C. Gourin, R. Faust and D. J. Terris (2007). 'A History of Robots: from Science Fiction to Surgical Robotics'. *Journal of Robotic Surgery* 1(2), 113–118.

Hoesslin, K. von (2016). 'The Economics of Piracy in South East Asia'. *The Global Initiative Against Transnational Organized Crime.*

Horowitz, M. C. (2006). *The Diffusion of Military Power: Causes and Consequences for International Politics.* Cambridge, MA: Harvard University.

Horowitz, M. C. (2010). *The Diffusion of Military Power: Causes and Consequences for International Politics.* Princeton: Princeton University Press.

Horowitz, M. C. (2014). 'The Looming Robotics Gap: America's Global Dominance in Military Technology Is Starting to Crumble'. *Foreign Policy.*

Horowitz, M. C. (2016). 'Public Opinion and the Politics of the Killer Robots Debate'. *Research and Politics* 3(1), 1–8.

Horowitz, M. C. (2016). 'The Ethics & Morality of Robotic Warfare: Assessing the Debate over Autonomous Weapons'. *Daedalus* 145(4), 25–36.

Horowitz, M. C. (2016). 'Why Words Matter: The Real World Consequences of Defining Autonomous Weapons Systmes'. *Temp. Int'l & Comp. Law Journal* 30, 85.

Horowitz, M. C. (2018). 'Artificial Intelligence, International Competition, and the Balance of Power'. *Texas National Security Review* 1, 3.

Horowitz, M. C. (2019). 'When Speed Kills: Lethal Autonomous Weapon Systems, Deterrence and Stability'. *Journal of Strategic Studies* 42(6), 764–788.

Horowitz, M. C., G. C. Allen, E. B. Kania and P. Scharre (2018). 'Strategic Competition in an Era of Artificial Intelligence'. *Artificial Intelligence and International Security*, Centre for a New American Security.

Hourihan, M. (2019). 'The FY 2020 Budget Request: Security R&D'. 23 April 2019, *American Association for the Advancement of Science.*

Hourihan, M. and D. Parkes (2016). 'Federal R&D Budget Trends: A Short Summary'. *Federal R&D Budget Overview*, AASS.

Huang, H. S. (2009). *An About-Face to the Future: The SAF's New Career Schemes.* S. Rajaratnam School of International Studies.

Huggler, J. (2015). 'Europe Faces a 'Real Threat' from Russia, Warns US Army Commander'. *The Telegraph.*

ICRC (2014). *Autonomous Weapon Systems: Technical, Military, Legal and Humanitarian Aspects.* Expert Meeting, Switzerland.

ICRC (2015). *International Humanitarian Law and the Challenges of Contemporary Armed Conflicts.* International Committee of the Red Cross.

ICRC (2016). 'Autonomous Weapon Systems Implications of Increasing Autonomy in the Critical Functions of Weapons'. *Expert Meeting.* Versoix, Switzerland.

ICRC (2016). 'Views of the International Committee of the Red Cross (ICRC) on Autonomous Weapon System'. *Convention on Certain Conventional Weapons (CCW) Meeting of Experts on Lethal Autonomous Weapons Systems (LAWS)*, Geneva.

Ikenberry, G. J. (2016). 'Between the Eagle and the Dragon: America, China, and Middle State Strategies in East Asia'. *Political Science Quarterly* 131(1), 9–43.

Indonesia, D. M. o. t. R. o (2015). *Defence White Paper.* Defence Ministry of the Republic of Indonesia.

Initiative, O. R. (2015). *Summary Report—The Ethics and Governance of Lethal Autonomous Weapons Systems: An International Public Opinion Poll.* Open Roboethics Initiative.

Institute, A.C. R. (2015). *South China Sea: What the Others Are Doing.* University of Technology Sydney.

Institute, F. o. L. (2015). *Autonomous Weapons: an Open Letter from Ai and Robotics Researchers.*

International, A. (2015). *Autonomous Weapons Systems: Five Key Human Rights Issues for Consideration.* London: Amnesty International.

International, A. (2017). *The Development of International Standards on the Export and Subsequent Use Of 'Armed or Strike-Enabled UAVs'.* London: Amnesty International.

International Institute for Strategic Studies. (2018). Chapter Six: Asia. In: *The Military Balance* J. Hackett (ed.) (Vol. 118, pp. 219–314). Routledge, London.

International Institute for Strategic Studies. (2019). Chapter Six: Asia. In: *The Military Balance*, *119*(1), 222–319. doi:10.1080/04597222.2018.1561032

IPSOS (2017). *Data for 2017 Campaign to Stop Killer Robots Survey.*

IPSOS (2019). *Six in Ten (61%) Respondents Across 26 Countries Oppose the Use of Lethal Autonomous Weapons Systems.*

Iswaran, S. (2019). 'Speech by Mr S Iswaran, Minister for Communications and Information and Minister-in-Charge of Cybersecurity'. Delivered at the *Total Defence Day Commemoration Event 2019* on 15 February 2019, Ministry of Communications and Information.

Ivanova, K., G. E. Gallasch and J. Jordans (2016). *Automated and Autonomous Systems for Combat Service Support: Scoping Study and Technology Prioritisation.* Edinburgh, SA: Defence Science and Technology Group.

Jamrisko, M. and H. Amin (2017). *Could Tech Relieve Singapore's Aging Woes?.* Bloomberg Technology.

Jane's Sentinel Security Assessment—Southeast Asia (2018). 'Singapore—Armed Forces'. Jane's by IHS Markit.

Jenks, C. (2016). 'The Distraction of Full Autonomy & the Need to Refocus the CCW LAWS Discussion on Critical Functions'. *Legal Studies Research. Paper*, SMU Dedman School of Law.

Jenne, N. J. (2017). 'Managing Territorial Disputes in Southeast Asia: Is There More than the South China Sea?' *Journal of Current Southeast Asian Affairs* 36(3), 35–61.

Kahn, P. W. (2002). 'The Paradox of Riskless Warfare'. *Philosophy and Public Policy Quarterly* 22(3), 2–7.

Kan, M. (2019). 'DeepMind's AI to Take on Human StarCraft II Players on Battle.net'. *PC Magazine Australia.*

Kania, E. B. (2017). 'Battlefield Singularity: Artificial Intelligence, Military Revolution, and China's Future Military Power'. *Center for a New American Security.*

Kania, E. B. (2018). 'China's Strategic Ambiguity and Shifting Approach to Lethal Autonomous Weapons Systems'. *Lawfare*, April 17, 2018, https://www.lawfareblog.com/chinas-strategic-ambiguity-and-shifting-approach-lethal-autonomous-weapons-systems.

Kania, E. B. (2018). 'China's AI Giants Can't Say No to the Party'. *Foreign Policy.*

Kania, E. B. (2019). 'Chinese Military Innovation in Artificial Intelligence'. 7 June 2019.

Kastan, B. (2013). 'Autonomous Weapons Systems: A Coming Legal "Singularity"?' *Journal of Law, Technology & Policy* 45(1), 45–82.

Keane, J. F. and S. S. Carr (2013). 'A Brief History of Early Unmanned Aircraft'. *Johns Hopkins APL Technical Digest* 32(3), 558–571.

Kearn, K. (2018). 'DoD Autonomy Roadmap: Autonomy Community of Interest'. *NDIA 19th Annual Science & Engineering Technology Conference.* Austin, Texas.

Klein, D. (2018). *Unmanned Systems & Robotics in the FY2019 Defense Budget.* Association For Unmanned Vehicle Systems International.

Kocak, D. (2013). 'Insurgencies, Border Clashes, and Security Dilemma--Unresolved Problems for ASEAN'. *Central European Journal of International and Security Studies* 7, 1.

Kraft, H. J. S. (2017). 'Great Power Dynamics and the Waning of ASEAN Centrality in Regional Security'. *Asian Politics and Policy* 9(4), 597–612.

Krepinevich, A. F. (1992). *The Military-Technical Revolution: A Preliminary Assessment.* Washington, DC: Center for Strategic and Budgetary Assessments.

Krepinevich, A. F. (1994). 'Cavalry to Computer: The Pattern of Military Revolutions'. *The National Interest* 37, 30–42.

Krishnan, A. (2009). 'Automating War: The Need for Regulation'. *Contemporary Security Policy* 30(1), 172–193.

Krishnan, A. (2013). *Killer Robots: Legality and Ethicality of Autonomous Weapons.* Ashgate Publishing Limited, Fanham.

Krisnamurthi, I. (2017). 'Statement by H.E. Ms. Ina H. Krisnamurthi Ambassador Deputy Permanent Representative of the Republic of Indonesia to the United Nations on behalf of the Non-Aligned Movement'. *The General Debate of the First Committee of the 72nd Session of the United Nations General Assembly.*

Kuik, C.-C. (2016). 'How Do Weaker States Hedge? Unpacking ASEAN States' Alignment Behavior towards China'. *Journal of Contemporary China* 25(100), 500–514.

Kwok Song Lee, J. (2015). 'The Limits of the ASEAN Regional Forum'. Master of Arts in Security Studies (Far East, Southeast Asia, The Pacific), Naval Postgraduate School.

Laboratory, D. S. a. T. (2018, 1 January). 'Guidance: Future Sensing and Situational Awareness Programme'. *DSTL's Work: Programmes and Facilities.* https://www.gov .uk/guidance/future-sensing-and-situational-awareness-programme.

Laksmana, E. A. (2014). 'The Hidden Challenges of Indonesia's Defence Modernisation'. *Indonesian Defence* 34(3), 17–19.

Laksmana, E. A. (2017). 'Threats and Civil–Military Relations: Explaining Singapore's "Trickle Down" Military Innovation'. *Defense and Security Analysis* 33(4), 347–365.

Laksmana, E. A. (2017). 'Pragmatic Equidistance: How Indonesia Manages Its Great Power Relations'. In: *China, The United States, and the Future of Southeast Asia* D. B. H. Denoon (ed.). New York: New York University Press, 113–135.

Laksmana, E. A. (2018). 'Are Military Assistance Programs Important for US–Indonesia Ties?'. 18 April 2018, *East Asia Forum.*

Laksmana, E. A. (2018). 'Is Southeast Asia's Military Modernization Driven by China? It's Not That Simple'. *Global Asia.*

Laksmana, E. A. (2018). 'Why Is Southeast Asia Rearming? An Empirical Assessment'. In: *U.S. Policy in Asia-Perspectives for the Future* R. Dossani and S. W. Harold (eds.). Santa Monica: RAND Corporation, 106–137.

Laksmana, E. A. (2019).'Reshuffling the Deck? Military Corporatism, Promotional Logjams and Post-Authoritarian Civil-Military Relations in Indonesia'. *Journal of Contemporary Asia*, 1–31.

Larsen, T. K. (2018). *Power and Arms: The Diffusion of Military Innovations and Technology.* University of Bergen. Department of Comparative Politics.

Le Thu, H. (2019). 'China's Dual Strategy of Coercion and Inducement towards ASEAN'. *The Pacific Review* 32(1), 20–36.

Lean, C. K. S. (2016). 'Speaking Notes'. *CCW Meeting of Informal Experts on Lethal Autonomous Weapon Systems*, Geneva.

Lebow, R. N. and B. Valentino (2009). 'Lost in Transition: A Critical Analysis of Power Transition Theory'. *International Relations* 23(3), 389–410.

Leiner, B. M., V. G. Cerf, D. D. Clark, R. E. Kahn, L. Kleinrock, D. C. Lynch, J. Postel, L. G. Roberts and S. Wolff (2009). 'A Brief History of the Internet'. *ACM Sigcomm Computer Communication Review* 39(5), 22–31.

Leone, D. (2019). 'South Korean Fighters Fired 300 Warning Shots at Russian A-50 AEW&C Aircraft'. *The National Interest.*

Lester, C. (2018). 'What Happens When Your Bomb-Defusing Robot Becomes a Weapon'. *The Atlantic.*

Libel, T. and E. Boulter (2015). 'Unmanned Aerial Vehicles in the Israel Defense Forces: A Precursor to a Military Robotic Revolution?' *The RUSI Journal* 160(2), 68–75.

Lin-Greenberg, E. (2016). 'So China Seized a U.S. Drone Submarine? Welcome to the Future of International Conflict'. *The Washington Post.*

Long, D. (2018). 'China Releases Video of 56-Boat Drone Swarm near Hong Kong'. *The Defense Post.*

Lubold, G. (2018). 'Russian Jet Fighter Buzzes U.S. Surveillance Plane Over Black Sea'. *The Wall Street Journal.*

Lucas, G. R., Jr (2014). 'Automated Warfare'. *Stanford Law and Policy Review* 25.

Mangosing, F. (2013). 'PH Army Displays Drones to Public'. 19 December 2013, *Inquirer.net.*

Mapp, W. (2014). 'Military Modernisation and Buildup in the Asia Pacific: The Case for Restraint'. *RSIS Monograph*, S, Rajaratnam School of International Studies.

Margulies, P. (2016). 'Making Autonomous Weapons Accountable: Command Responsibility for Computer-Guided Lethal Force in Armed Conflicts'. In: *Research Handbook on Remote Warfare* Jens David Ohlin (ed.). Cheltenham: Edward Elgar Press, 405–442.

Markit, I. (2016). *$20 Billion Defence Budget Boom in Indonesia.* IHS Markit.

Markowski, S., P. Hall and R. Wylie (2009). *Defence Procurement and Industry Policy: A Small Country Perspective.* Taylor & Francis.

Martin, C. (2015). 'A Means-Methods Paradox and the Legality of Drone Strikes in Armed Conflict'. *The International Journal of Human Rights* 19(2), 142–175.

Marwati (2017). *UGM - Indonesian Defense University Collaborate in Defence Technology Development.* Universitas Gadjah Mada.

Matthews, R. and N. Z. Yan (2007). 'Small Country 'Total Defence': A Case Study of Singapore'. *Defence Studies* 7(3), 376–395.

McKitrick, J., J. Blackwell, F. Littlepage, G. Kraus, R. Blanchfield and D. Hill (1998). 'The Revolution in Military Affairs'. In: *Battlefield of the Future: 21st Century Warfare Issues* B. R. Schneider and L. E. Grinter (ed.). Maxwell Air Force Base, AL: Air University Press, 65–102.

Mearsheimer, J. J. (2004). 'Hitler and the Blitzkrieg Strategy'. In: *The Use of Force: Military Power and International Politics* R. J. Art and K. N. Waltz (eds). London: Rowman & Littlefield, 155–166.

Mearsheimer, J. J., (ed.) (2013). 'Structural Realism'. In: *International Relations Theories: Discipline and Diversity.* Oxford: Oxford University Press, 51–67.

Medcalf, R. (2014). 'Asia's "Cold Peace": China and India's Delicate Diplomatic Dance'. *Brookings Institution Opinions.* https://www.brookings.edu/opinions/asias-cold-peace-china-and-indias-delicate-diplomatic-dance/.

Meeting, A. D. M. (2017). 'Concept Paper on the Guidelines for Maritime Interaction'.

Meeting, A. D. M. (2018). ''Guidelines for Air Military Encounters'. *12th ASEAN Defence Ministers' Meeting.*

Meeting, A. D. M. (2019). 'ASEAN Defence Ministers' Meeting (ADMM) Guidelines for Maritime Interaction'. *13th ASEAN Defence Ministers' Meeting.*

Mehta, A. (2016). 'White House Rolls Out Armed Drone Declaration'. 5 October 2016, *Defense News.*

Mehta, A. (2018). 'Experiment over: Pentagon's Tech Hub Gets a Vote of Confidence'. 9 August 2018, *Defense News.*

Metz, S. and J. Kievit (1994). 'The Revolution in Military Affairs and Conflict Short of War'. Army War College Strategic Studies Institute, Carlisle Barracks PA; Metz, S. and J.

Kievit (1995). 'Strategy and the Revolution in Military Affairs: From Theory to Policy'. Diane Publishing; Black, J. (2004). 'Rethinking military history'. Psychology Press.

Miller, P. M. (2006). *Mini, Micro, and Swarming Unmanned Aerial Vehicles: A Baseline Study*. Washington, DC: Federal Research Division, Library of Congress.

Minister Iswaran is also Singapore's Minister-in-charge of Cyber Security.

Ministry of Defence, S. (2018). 'Fact Sheet: Unmanned Watch Towers—Enhancing the SAF's Protection of Installation Operations'. 2 March 2018. https://www.mindef.gov. sg/web/portal/mindef/news-and-events/latest-releases/article-detail/2018/march/02m ar18_fs3.

Ministry of Research, T. a. H. E. o. t. R. o. I. (n.d.). 'Statistics'. https://international.ristekdi kti.go.id/statistics/.

Mizokami, K. (2017). 'Kaboom! Russian Drone With Thermite Grenade Blows Up a Billion Dollars of Ukrainian Ammo'. *Popular Mechanics*.

Moncada, S. (2018). 'Baku Declaration of the 18th Midterm Ministerial Meeting of the Non-Aligned Movement'. Baku, Republic of Azerbaijan.

Moon, C.-i. and J.-Y. Lee (2008). 'The Revolution in Military Affairs and the Defence Industry in South Korea'. *Security Challenges* 4(4), 117–134.

Morris, I. (2014). *War: What Is It Good For? The Role of Conflict in Civlisation, from Primates to Robots*. London: Profile Books Ltd.

Morris, L. J. and G. Persi Paoli (2018). *A Preliminary Assessment of Indonesia's Maritime Security Threats and Capabilities*. RAND Corporation.

Mugg, J., Z. Hawkins and J. Coyne (2016). *Australian Border Security and Unmanned Maritime Vehicles*. Border Security Program. Australian Strategic Policy Institute.

Mugg, J., Zoe Hawkins and John Coyne (2016). 'AWD Combat System: An Upgrade for the Aegis'. *ASPI Strategic Insights*, July 2016.

Mukhtar, A., L. Xia and T. B. Tang (2015). 'Vehicle Detection Techniques for Collision Avoidance Systems: A Review'. *IEEE Transactions on Intelligent Transportation Systems* 16(5), 2318–2338.

Mullens, K. D., E. B. Pacis, S. B. Stancliff, A. B. Burmeister and T. A. Denewiler (2003). *An Automated UAV Mission System*. DTIC Document.

Nathan, L. (2006). 'Domestic Instability and Security Communities'. *European Journal of International Relations* 12(2), 275–299.

Newcome, L. R. (2004). *Unmanned Aviation: A Brief History of Unmanned Aerial Vehicles*. American Institute of Aeronautics and Astronautics, Reston Virginia.

News Desk (2018). 'Indonesia Leads Asia-Pacific in AI Implementation, Study Shows'. 5 September 2018, *The Jakarta Post*.

News, S. (2017). *Aerial Ghosts: Russia's Autonomous 5th Gen Su-57 to Dominate the Skies*, 14 August 2017, from https://sputniknews.com/russia/201708141056451784-russia-plane-computer/.

Newsroom, A. S. (2017). 'NVIDIA Partners BINUS for Indonesia's First AI Research Center'. *Asian Scientist*.

Non-Aligned Movement (2016). 'Final Document'. *17th Summit of Heads of State and Government of the Non-Aligned Movement*, Island of Margarita, Bolivarian Republic of Venezuela.

Nupus, A. (2018). 'Orders for 100 Turkish-Indonesian Medium Battle Tank'. 7 September 2018, *Defence Aerospace*.

Nurkin, T., K. Bedard, J. Clad, C. Scott and J. Grevatt (2018). 'China's Advanced Weapons Systems'. *Prepared for the U.S.-China Economic and Security Review Commission*, Jane's by IHS Markit.

Nye Jr, J. S. (2010). 'The Futures of American Power-Dominance and Decline in Perspective'. *Foreign Affairs* 89.

Nye, J. S. (2011). 'The Rise and Fall of Great Powers'. *War and Peace in the 20th Century and Beyond*, World Scientific, 121–144.

OECD (2018). *Economic Outlook for Southeast Asia, China and India 2018: Fostering Growth Through Digitalisation.* Paris: OECD Publishing.

Office of the Secretary of Defense. (2015). 'Military and Security Developments Involving the People's Republic of China 2015', *Annual Report to Congress.*

Online, T. (2016). *SAF Looks to Artificial Intelligence to Gain Punch.* Singapore Government.

Organisation, D. I. (2018). *Defence Economic Trends in the Asia-Pacific 2018.* Australia: Australian Department of Defence.

Otto, R. P. (2016). *Small Unmanned Aircraft Systems (SUAS) Flight Plan: 2016–2036. Bridging the Gap Between Tactical and Strategic.* Washington, DC: United States Air Force.

Panda, A. (2018). 'Singapore: A Small Asian Heavyweight'. *Backgrounder*, Council on Foreign Relations.

Parameswaran, P. (2016). 'Vietnam Now World's Eighth Largest Arms Importer'. 23 February 2016, *The Diplomat.* http://thediplomat.com/2016/02/vietnam-now-wo rlds-eighth-largest-arms-importer/.

Parameswaran, P. (2017). 'Indonesia's War on Illegal Fishing Nets New China Vessel'. 06 December 2017, *The Diplomat.*

Parameswaran, P. (2018). 'Indonesia Spotlights Defense Industry Challenge Under Jokowi'. 10 March 2018, *The Diplomat.*

Parameswaran, P. (2018). 'Thailand's Navy Gets a Boost with Five New Patrol Vessels'. 21 February 2018, *The Diplomat.*

Parameswaran, P. (2018). 'What's in the New Indonesia Defense Industry Financing Pact?'. 10 May 2018, *The Diplomat.*

Parameswaran, P. (2019). 'What's in the New Indonesia South China Sea Base Hype?'. 11 March 2019, *The Diplomat.* https://thediplomat.com/2019/01/whats-in-the-new-i ndonesia-south-china-sea-base-hype/.

Parameswaran, P. (2019). 'What's Next for the Indonesia-Russia Fighter Jet Deal in 2019?' 11 June 2019, *The Diplomat.* https://thediplomat.com/2019/06/whats-next-for-the-indo nesia-russia-fighter-jet-deal-in-2019/.

Parameswaran, P. (2019). 'What's in Indonesia's New Natuna Fishing Zone in the South China Sea?' 23 February 2019, *The Diplomat.*

Parkins, S. (2015). 'Killer Robots: The Soldiers That Never Sleep'. 16 July 2015, *BBC NEWS.*

Parmar, T. (2015). 'Drones in Southeast Asia'. *Centre for the Study of the Drone.* https:// dronecenter.bard.edu/drones-in-southeast-asia/.

Pengembangan, B. P. d. (2018). 'Puslitbang Iptekhan'. https://www.kemhan.go.id/balit bang/tupoksi-iptekhan.

Pengembangan, B. P. d. (2018). 'Uji Fungsi Rancang Bangun Sistem Persenjataan Sentry Gun Pada Ranpur'. *Kementerian Pertahanan Republik Indonesia.*

Pengembangan, B. P. d. (2018, 13 August). 'Hasil Litbang Puslitbang Iptekhan Ta. 2018'. https://www.kemhan.go.id/balitbang/2018/08/13/hasil-litbang-puslitbang-iptekhan-ta-2018.html.

Perlez, J. and M. Rosenberg (2016). 'China Agrees to Return Seized Drone, Ending Standoff, Pentagon Says'. *New York Times.*

Permanent Mission of the Kingdom of Thailand to the United Nations (2017). 'Statement Delivered by H.E. Mr. Virachai Plasai, Ambassador and Permanent Representative of the Kingdom of Thailand to the United Nations at the General Debate of the First Committee'. 2nd Meeting of the First Committee, Seventy-second Session of the United Nations General Assembly, United Nations General Assembly.

Pertahanan, U. (2016). 'Vision and Mission'. http://www.idu.ac.id/profil/visi-misi.

Picard, M. (2019). 'Weaponized AI in Southeast Asia: In Sight yet Out of Mind'. 06 July 2019, *The Diplomat*. https://thediplomat.com/2019/07/weaponized-ai-in-southeast-asia-in-sight-yet-out-of-mind/.

Pilloud, C., Y. Sandoz, C. Swinarski and B. Zimmermann (1987). *Commentary on the Additional Protocols: Of 8 June 1977 to the Geneva Conventions of 12 August 1949*. Leiden: Martinus Nijhoff Publishers.

Polmar, N. (2013). 'The Pioneering Pioneer'. *Naval History* 27(5), 14.

Purbrick, M. (2018). 'Pirates of the South China Seas'. *Asian Affairs* 49(1), 11–26.

Rabasa, A. and J. Haseman (2002). *The Military and Democracy in Indonesia: Challenges, Politics, and Power*. RAND Corporation.

Rahakundini, C. and A. Prasetia (2016). 'A Perspective on Indonesia'. *Proliferated Drones*. Center for a New American Security. Retrieved from http://drones.cnas.org/reports/a-perspective-on-indonesia/#1463433864196-3bb6096f-8aaf

Rahmat, R. (2018). 'Indonesian Navy to Establish New Unmanned Aviation Squadron'. *Jane's Navy International*. IHS Markit.

Rahmat, R. (2019). 'Tensions between Malaysia, Singapore Re-Escalate after Minister's 'Intrusion''. *Jane's Navy International*, Jane's 360.

Raska, M. (2015). *Military Innovation in Small States: Creating a Reverse Asymmetry*. London, UK: Routledge.

Raska, M. (2019). 'How Will SAF Look Like after Its Next Incarnation?' *Today*.

Raska, M. (n.d.). 'Strategic Transformation and Military Modernization in the Asia-Pacific Region'. *Draft Paper*.

Ray, T. (2018). 'Beyond the 'Lethal' in Lethal Autonomous Weapons: Applications of LAWS in Theatres of Conflict for Middle Powers'. *ORF Occasional Paper*, Observer Research Foundation.

Raymond, G. V. (2017). 'Naval Modernization in Southeast Asia: Under the Shadow of Army Dominance?' *Contemporary Southeast Asia* 39(1), 149–177.

Resende-Santos, J. (1996). 'Anarchy and the Emulation of Military Systems: Military Organization and Technology in South America, 1870–1930'. *Security Studies* 5(3), 193–260.

Reuters (2014). 'New Focus on Arms Industry Expansion in Southeast Asia'. 12 August 2014, *New York Times*.

Reuters (2016). 'Indonesian Navy Fires on Chinese Fishing Boat, Injuring One, Beijing Claims'. 20 June 2016, *The Guardian*.

Robots, C. t. S. K. (n.d.). 'The Problem'. Retrieved 29 August 2017, from http://www.stopkillerrobots.org/the-problem/.

Robots, C. t. S. K. (2017). *Country Views on Killer Robots: 11 October 2017*.

Robots, C. t. S. K. (2018). 'Members'. https://www.stopkillerrobots.org/members/.

Robots, C. t. S. K. (2019). 'Global Poll Shows 61% Oppose Killer Robots'. https://www.stopkillerrobots.org/2019/01/global-poll-61-oppose-killer-robots/.

Roggeveen, S. (2018). 'Combat Drones: Australia's Uncertain Future'. *The Interpreter*. https://www.lowyinstitute.org/the-interpreter/australia-combat-drone-future.

Rosen, S. P. (1988). 'New Ways of War: Understanding Military Innovation'. *International Security* 13(1), 134–168.

Rosen, S. P. (1991). *Innovation and the Modern Military: Winning the Next War.* Ithaca, NY: Cornell University Press.

Rosser, A. (2018). 'Improving Education Quality in Indonesia Is No Easy Task'. *The Interpreter.* https://www.lowyinstitute.org/the-interpreter/debate/education-indonesia.

Rosser, A. (2018). 'Beyond Access: Making Indonesia's Education System Work'. *Analyses, Lowy Institute.*

RSIS (2014). 'Revitalizing Indonesia's Defence Industrial Base: Agenda for Future Action'. *RSIS Indonesia Program.* S. Rajaratnam School of International Studies.

Ryan, M. (2018). *Human-Machine Teaming for Future Ground Forces.* Center for Strategic and Budgetary Assessments.

Saitou, K. (2012). 'The Diffusion of Military Power: Causes and Consequences for International Politics by Michael C. Horowitz'. *Interfaculty* 3.

Sample, I. (2017). 'Ban on Killer Robots Urgently Needed, Say Scientists'. 13 November 2017, *The Guardian.* https://www.theguardian.com/science/2017/nov/13/ban-on-killer-robots-urgently-needed-say-scientists.

Santikajaya, A. (2016). 'Walking the Middle Path: The Characteristics of Indonesia's Rise'. *International Journal* 71(4), 563–586.

Sauer, F. (2016). 'Stopping 'Killer Robots': Why Now Is the Time to Ban Autonomous Weapon Systems'. https://www.armscontrol.org/print/7713.

Sayler, K. (2015). 'A World of Proliferated Drones'. *Center for a New American Security.*

Scharre, P. (2016). 'Autonomous Legal Reasoning?: Legal and Ethical Issues in the Technologies of Conflict: Centaur Warfighting: The False Choice of Humans vs. Automation'. *Temple International and Comparative Law Journal* 30, 151–177.

Scharre, P. (2018). *Army of None: Autonomous Weapons and the Future of War.* WW Norton & Company.

Schmid, J. (2018). 'The Determinants of Military Technology Innovation and Diffusion'. Doctor of Philosophy in International Affairs, Science, and Technology, Georgia Institute of Technology.

Schmitt, M. (2013). 'Autonomous Weapon Systems and International Humanitarian Law: A Reply to the Critics'. *Harvard National Security Journal Features.*

Schmitt, M. and J. Thurnher (2012). 'Out of the Loop: Autonomous Weapon Systems and the Law of Armed Conflict'. *Harvard National Security Journal* 4, 231–281.

Schneider, J. (2019). 'The Capability/Vulnerability Paradox and Military Revolutions: Implications for Computing, Cyber, and the Onset of War'. *Journal of Strategic Studies* 42(6), 841–863.

Schreer, B. (2015). 'Garuda Rising?: Indonesia's Arduous Process of Military Change'. In: *Security: Strategy and Military Change in the 21st Century* J. I. Bekkevold, I. Bowers and M. Raska (eds.). Routledge, 55–69.

Schulzke, M. (2018). 'Drone Proliferation and the Challenge of Regulating Dual-Use Technologies'. *International Studies Review* 21(3), 497–517.

Searight, A. (2018). 'ADMM-Plus: The Promise and Pitfalls of an ASEAN-Led Security Forum'. *Centre for Strategic & International Studies.* https://www.csis.org/analysis/admm-plus-promise-and-pitfalls-asean-led-security-forum.

Sebastian, L. C. and I. Gindarsah (2011). 'Assessing 12-Year Military Reform in Indonesia: Major Strategic Gaps for the Next Stage of Reform'. *RSIS Working Paper.* S. Rajaratnam School of International Studies, 227.

Sebastian, L. C. and I. Gindarsah (2013). 'Assessing Military Reform in Indonesia'. *Defense and Security Analysis* 29(4), 293–307.

Sebastian, L. C., E. A. Syailendra and K. I. Marzuki (2018). 'Civil-Military Relations in Indonesia after the Reform Period'. *Asia Policy* 25(3), 49–78.

Sechser, T. S., N. Narang and C. Talmadge (2019).'Emerging Technologies and Strategic Stability in Peacetime, Crisis, and War'. *Journal of Strategic Studies* 42(6), 727–735.

Secretary, O. o. t. P. (2015). *Joint Statement by the United States of America and the Republic of Indonesia.* The White House.

Sehrawat, V. (2017). 'Autonomous Weapon System: Law of Armed Conflict (LOAC) and Other Legal Challenges'. *Computer Law and Security Review* 33(1), 38–56.

Setiawan, R. (2018). 'Indonesia Cyber Security: Urgency to Establish Cyber Army in the Middle of Global Terrorist Threat'. *Journal of Islamic World and Politics* 2(1), 157–173.

Shachtman, N. (2007). 'Robot Cannon Kills 9, Wounds 14'. 18 October 2007, *Wired.*

Sharkey, N. (2010). 'Saying 'No!'to Lethal Autonomous Targeting'. *Journal of Military Ethics* 9(4), 369–383.

Sharkey, N. (2011). 'The Automation and Proliferation of Military Drones and the Protection of Civilians'. *Law, Innovation and Technology* 3(2), 229–240.

Sharkey, N. (2017). 'Why Robots Should Not Be Delegated with the Decision to Kill'. *Connection Science* 29(2), 177–186.

Shewan, D. (2017). 'Robots Will Destroy Our Jobs—And We're Not Ready for It'. 11 January 2017, *The Guardian.*

Shoop, B., M. Johnston, R. Goehring, J. Moneyhun and B. Skibba (2006). *Mobile Detection Assessment and Response Systems (MDARS): A Force Protection, Physical Security Operational Success.* San Diego: Space and Naval Warfare Systems Center.

Silverstein, A. B. (2013). 'Revolutions in Military Affairs: A Theory on First-Mover Advantage'. 01 April 2013, *CUREJ: College Undergraduate Research Electronic Journal*, University of Pennsylvania. http://repository.upenn.edu/curej/169.

Silviana, C. and E. Danubrata (2015). 'Rising SE Asia Defense Spending to Spur Sales for Indonesia's Pindad'. 17 March 2015, *Reuters.*

Simon, S. W. (2012). 'Conflict and Diplomacy in the South China Sea'. *Asian Survey* 52(6), 995–1018.

Singer, P. W. (2009). *Wired for War: The Robotics Revolution and Conflict in the 21st Century.* Penguin Publishing Group.

Singer, P. W. and A. Cole (2015). *Ghost Fleet: A Novel of the Next World War.* Houghton Mifflin Harcourt.

Solution, W. I. T. (2018). *Indonesia Trade at a Glance : Most Recent Values.* The World Bank, Washington D.C.

Sparrow, R. (2015). 'Twenty Seconds to Comply: Autonomous Weapons Systems and the Recognition of Surrender'. Int'l L. Stud. Ser. US Naval War Col 91, 699.

Sparrow, R. (2016). 'Robots and Respect: Assessing the Case against Autonomous Weapon Systems'. *Ethics and International Affairs* 30(1), 93–116.

Staff, J. s. E. (2017). 'Global Defence Exports Expected to Decline for First Time Ever'. *Jane's Aerospace Defense and Security Blog.* http://blog.ihs.com/global-defence-exports-expected-to-decline-for-first-time-ever.

Staff, J. s. E. (2018). 'Singapore Airshow 2018 Preview'. *Aerospace, Defense & Security.* IHS Markit.

Starke, P. (2013). 'Qualitative Methods for the Study of Policy Diffusion: Challenges and Avaliable Solutions'. *Policy Studies Journal* 42(4), 561–582.

Steuter, E. and D. Wills (2009). *At War with Metaphor: Media, Propaganda, and Racism in the War on Terror*. Lexington Books.

Stevenson, B., N. Sharkey, N. Marsh and R. Crootof (2015). 'Special Session 10: How to Regulate Autonomous Weapon Systems'. Paper presented at the 2015 EU Non-Proliferation and Disarmament Conference, International Institute for Strategic Studies, Brussels.

Stohl, R. (2015). 'Exercising Restraint? The New US Rules for Drone Transfers'. *Arms Control Today* 45(4), 20.

Strawser, B. J. (2010). 'Moral Predators: The Duty to Employ Uninhabited Aerial Vehicles'. *Journal of Military Ethics* 9(4), 342–368.

Studies, I. I. f. S. (2018). 'Chapter Six: Asia'. In: *The Military Balance* J. Hackett (ed.). Routledge, 219–314.

Studies, I. I. f. S. (2019). 'Chapter Six: Asia'. In: *The Military Balance* J. Hackett (ed.). Routledge, 222–319.

Studies, S. R. S. o. I. (2014). 'Indonesia's Emerging Defence Economy: The Defence Industry Law and Its Implications'. *Indonesia Programme*, S. Rajaratnam School of International Studies.

Sukma, R. (2012). 'Indonesia and the Emerging Sino-US Rivalry in Southeast Asia'. *The Geopolitics of Southeast Asia*, LSE IDEAS Special Report, London: London School of Economics and Political Science.

Syailendra, E. A. (2017). 'A Nonbalancing Act: Explaining Indonesia's Failure to Balance Against the Chinese Threat'. *Asian Security* 13(3), 237–255.

Systems, H. (2018). 'Core Competencies'. http://www.hanwhasystems.com/eng/company/competencies.do.

Systems, H. L. (2018). 'Products at a Glance'. http://www.hanwhalandsystems.com/products/overview.

Talmadge, C. (2019).'Emerging Technology and Intra-War Escalation Risks: Evidence from the Cold War, Implications for Today'. *Journal of Strategic Studies* 42(6), 864–887.

Tan, A. (2019). 'SG Budget 2019: Home Team Science & Tech Agency to Be Set Up by End 2019'. 18 February 2019, *Vulcan Post*.

Tan, A. T. H. (2013). 'Singapore's Defence Industry: Its Development and Prospects'. *Security Challenges* 9(1), 63–85.

Tan, F. W.-S. and P. B. Lew (2017). 'The Role of the Singapore Armed Forces in Forging National Values, Image, and Identity'. *Military Review* 97(2), 8–16.

Tan, S. S. (2015). 'Mailed Fists and Velvet Gloves: The Relevance of Smart Power to Singapore's Evolving Defence and Foreign Policy'. *Journal of Strategic Studies* 38(3), 332–358.

Tan, S. S. (2017). 'A Tale of Two Institutions: The ARF, ADMM-Plus and Security Regionalism in the Asia Pacific'. *Contemporary Southeast Asia* 39(2), 259–264.

Tang, S. M. (2016). 'ASEAN and the ADMM-Plus: Balancing between Strategic Imperatives and Functionality'. *Asia Policy* 22(1), 76–82.

Tang, S.-M. (2018). 'ASEAN's Tough Balancing Act'. *Asia Policy* 25(4), 48–52.

Tao, A. L. (2018). 'Indonesia Leads ASEAN Region in AI Adoption'. ComputerWeekly.com.

Taylor, C. (2011). *Military Balance in Southeast Asia*. United Kingdom: House of Commons Library.

Thayer, C. A. (2018). 'Force Modernization: Vietnam'. *Southeast Asian Affairs* 1, 429–444.

Tian, N., A. Fleurant, A. Kuimova, P. D. Wezeman and S. T. Wezeman (2018). 'Trends in World Military Expenditure, May 2017'. *Sipri Fact Sheet*, Stockholm International Peace Research Institute, Stockholm.

Tjin-Kai, O. (2012). 'Interpreting Recent Military Modernizations in Southeast Asia: Cause for Alarm or Business As Usual?' *Pointer, Journal of the Singapore Armed Forces* 38(1), 13–31.

Tjin-Kai, O. (2012). 'Interpreting Recent Military Modernizations in Southeast Asia: Cause for Alarm or Business As Usual?' *Pointer, Journal of the Singapore Armed Forces*.

Trade, D. o. F. A. a. (2016). 'East Asia Summit (EAS)'. Retrieved 17 March 2017.

Trade, D. o. F. A. a. (2017). 'East Asia Summit Factsheet'. Retrieved 17 March 2017.

Tucker, P. (2018). 'This Stealthy Drone May Be The Future of Russian Fighter Jets'. 23 July 2018, *Defense One*.

Turner, K. B. (2016). 'Lethal Autonomous Weapons Systems: The Case For International Prohibition'. Masters of Science, Missouri State University.

Udoshi, R. (2019). 'MAKS 2019: Russia unveils export variant of Su-57'. 30 August 2019, *Jane's Defence Weekly*, IHS Jane's.

UNARM (2018). 'Military Spending'. United Nations.

Unknown (2013). 'Frost & Sullivan: Israel is the World's Largest Exporter of Unmanned Aircraft'. *IHLS Janes*.

Unknown (2016). 'The Laboratory for the Development of Optical Devices of the New Generation Was Created With the Support of the Foundation for Advanced Studies'. The Rare Earth Magazine.

Unknown (2018). 'ST Kinetics Unveils New Weaponised Probot UGV Unmanned Ground Vehicle'. *Arms Recognition*. https://www.armyrecognition.com/singapore_airshow:2018_latest_news/st_kinetics_unveils_new:weaponised_probot_ugv_unmanned_ground_vehicle.html.

Unknown (2018). 'PT PINDAD Conducted a Successful Live-Firing Demo of Its Latest "Medium Tank"'. *Asian Military Review*.

Unknown (2018). 'Russia's Su-57 Plane Tests Onboard Systems for 6th-Generation Fighter Jet'. *TASS—Russian News Agency*.

Unknown (2018). 'Russia's 'Syria Tested' Robotic Vehicle Shows Off Its Firepower'. *RT*.

Unlisted (2018). 'Weapons Manufacturer Pindad Diversifies Businesses'. *The Jakarta Post*, Jakarta.

Unmanned Systems Research Group (2018). 'Research Project List'. 7 May 2018. http://unmanned.kaist.ac.kr/project.htm.

US State Department. (2016). *Joint Declaration for the Export and Subsequent Use of Armed or Strike-Enabled Unmanned Aerial Vehicles (UAVs)*, Bureau of Public Affairs, October 28, 2016, https://2009-2017.state.gov/r/pa/prs/ps/2016/10/262811.htm

Valenti, A. (2017). 'Frigate or Destroyer?'. 25 April 2017, *Asian Military Review*.

Venezuela, B. R. o (2018). 'General Principles on Lethal Autonomous Weapons Systems: Submitted by the Bolivarian Republic of Venezuela on behalf of the Non-Aligned Movement (NAM) and Other States Parties to the Convention on Certain Conventional Weapons (CCW)'. Group of Governmental Experts of the High Contracting Parties to the Convention on Prohibitions or Restrictions on the Use of Certain Conventional Weapons Which May Be Deemed to Be Excessively Injurious or to Have Indiscriminate Effects. Geneva: United Nations.

Vickers, M. G. (2010). 'The Structure of Military Revolutions'. Doctor of Philosophy, Johns Hopkins University.

Vickers, M. G. and R. C. Martinage (2004). *The Revolution in War*. The Center for Strategic and Budgetary Assessments.

Vogel, R. (2010). 'Drone Warfare and the Laws of Armed Conflict'. *Denver Journal of International Law and Policy* 45(1), 45–82.

Volpe, T. A. (2019).'Dual-Use Distinguishability: How 3D-Printing Shapes the Security Dilemma for Nuclear Programs'. *Journal of Strategic Studies* 42(6), 814–840.

Wagner, M. (2014). 'The Dehumanization of International Humanitarian Law: Legal, Ethical, and Political Implications of Autonomous Weapon Systems'. *Vand. Journal of Transnational Law* 47, 1371.

Walsh, J. I. (2015). 'Political Accountability and Autonomous Weapons'. *Research and Politics* 2(4), 4.

Warrick, J. (2017). 'Use of Weaponized Drones by ISIS Spurs Terrorism Fears'. 21 February 2017, *Washington Post*.

Wassmuth, D. and D. Blair (2018). 'Loyal Wingman, Flocking, and Swarming: New Models of Distributed Airpower'. *War on the Rocks*. https://warontherocks.com/2018/02/loyal-wingman-flocking-swarming-new-models-distributed-airpower/.

Watch, H. R. (2019, 19 August). ''Killer Robots:' Russia, US Oppose Treaty Negotiations: New Law Needed to Retain Meaningful Human Control Over the Use of Force'. Retrieved 20 August 2019, from https://www.hrw.org/news/2019/08/19/killer-robots-russia-us-oppose-treaty-negotiations.

Wei, T. T. (2019). 'NTU and Volvo Launch World's First Full-Sized Driverless Electric Bus for Trial'. 05 March 2019, *The Straits Times*.

Wezeman, P. D., A. Fleurant, A. Kuimova, N. Tian and S. T. Wezeman (2018). 'Trends in International Arms Transfers, March 2017'. *SIPRI Fact Sheet*, Stockholm International Peace Research Institute.

Whittle, R. (2014). *Predator: The Secret Origins of the Drone Revolution*. Henry Holt and Company.

Wijaya, L. (2018). 'The Rise of Indonesian Nationalism in Response to Illegal Fishing'. 25 January 2018, Retrieved 03 March 2018, from https://theconversation.com/the-rise-of-indonesian-nationalism-in-response-to-illegal-fishing-86947.

Williams, H. (2017). 'Rafael Launches Spike Missiles from Protector USV'. 08 March 2017, *IHL Jane's International Defence Review*.

Williams, H. (2019).'Asymmetric Arms Control and Strategic Stability: Scenarios for Limiting Hypersonic Glide Vehicles'. *Journal of Strategic Studies* 42(6), 789–813.

Williamson, G. and I. Palmer (2012). *German E-Boats 1939–45*. Bloomsbury Publishing.

Wong, K. (2018). 'ST Kinetics Pursues Weaponised Multirotor UAV Development'. 6 March 2018, *Jane's International Defence Review*, IHS Jane's.

Wong, K. (2019). 'Singapore Outlines Next-Generation Armed Forces in Latest Transformation Roadmap'. 5 March 2019, *Jane's Defence Weekly*, IHS Markit.

Work, R. O. and G. Grant (2019). *Beating the Americans at Their Own Game: An Offset Strategy with Chinese Characteristics*. Centre for a New American Security.

Wyatt, A. and J. Galliott (2018). 'Closing the Capability Gap: ASEAN Military Modernization during the Dawn of Autonomous Weapon Systems'. *Asian Security*, 1–20.

Wyrtki, K. (1961). 'Physical Oceanography of the Southeast Asian Waters: Scientific Results of Marine Investigations of the South China Sea and the Gulf of Thailand 1959–1961'. *NAGA Report*, University of California.

Yong, J. R. L. Y. (2017). 'Why Keep Changing? Explaining The Evolution Of Singapore's Military Strategy Since Independence'. Master Of Arts In Security Studies (Far East, Southeast Asia, The Pacific), Naval Postgraduate School.

Yulisman, L. and N. A. M. Salleh (2019). 'Decision Time for Prabowo after Losing Indonesia Court Fight'. 29 July 2019, *The Straits Times*.

Zhang, F. (2018). 'Is Southeast Asia Really Balancing against China?' *The Washington Quarterly* 41(3), 191–204.

Zhang, L. M. (2018). 'Parliament: Defence Spending to Remain Steady Even as Other Countries Spend More on Wide-Ranging Security Threats, says Ng Eng Hen'. 2 March 2018, *The Straits Times*.

Zhou, L. (2016). 'China's Foreign Ministry Joins War of Words against Singapore over South China Sea Dispute'. 27 September 2016, *South China Morning Post*.

Index

For Product Safety Concerns and Information please contact our EU
representative GPSR@taylorandfrancis.com
Taylor & Francis Verlag GmbH, Kaufingerstraße 24, 80331 München, Germany

www.ingramcontent.com/pod-product-compliance
Lightning Source LLC
Chambersburg PA
CBHW060448240326
41598CB00088B/4074

9 781032 001555